GPS Principles and Applications

GPS Principles and Applications

Dr. A. Ganesh
R. Narayanakumar

SATISH SERIAL PUBLISHING HOUSE
403, Express Tower, Commercial Complex, Azadpur, Delhi - 110033 (INDIA)
Phone : 011-27672469, 27672852 Fax : 91-11-27672046
E-mail : ssp@ndf.vsnl.net.in, hkjain@satishserial.com
Website : www.satishserial.com

Published by :

SATISH SERIAL PUBLISHING HOUSE

403, Express Tower, Commercial Complex, Azadpur, Delhi-110033 (INDIA)
Phone : 011-27672852 Fax : 91-11-27672046
E-mail : info@satishserial.com, hkjain1975@yahoo.com

Reprint, 2019

ISBN 81-89304-24-0

Composed, Designed & Printed in India

Content

Preface

The advent of the Global Positioning System (GPS) has irreversibly transformed the landscape of the world. Developed by the U.S. military, GPS is a constellation of satellites that transmit their signals to receivers all over the world. GPS enables geodetic positioning to be accomplished without having to physically see or measure distances between survey points. Using GPS, a survey that once took days or weeks to complete can now be done in a few hours at a much lower cost. GPS has not only revolutionized the traditional civilian navigation, surveying, and mapping professions, but has spawned numesrous new applications in various sectors. Keeping the importance of GPS technology, an attempt has been made to study the principles of Global Positioning System.

The intellectual content of this book gestated over the years as we often tried to explain GPS as a science to geoscientists, and describe it as a process to the GIS students. In both the cases, we were forced to examine and strengthen our arguments. This book is directed at both these realms on geosciences, which we continue to straddle.

A Ganesh
R Narayanakumar

Chapter - 1

INTRODUCTION TO GPS TECHNOLOGY

Mankind has been trying to figure out a dependable way to know where they are, and to guide them to where they want to go and get back again. At the centre of any spatial tasks is the problem of location: where are you and where do you want to be? It might be as trivial as wanting to know where to get a square meal when it's late, or as important as letting rescue services know where you are when you're caught out of fire in an unexpected situation. Historically, the problem of location has been solved using local references. Be it your memory, a road map or a topographical map, it's up to you to figure out where you are by looking for reference points and then navigating to where it is you want to be. Of course if you're in a strange area or local landmarks are difficult to see, the problems of location can be enormous. But now, the problem of 'where am I?' has been effectively solved with the Global Positioning System.

The Global Positioning System (GPS) is a location system based on a constellation of 24 satellites orbiting the Earth at altitudes of approximately 20,200 km. GPS was developed by the United States Department of Defense (DOD), for its tremendous application as a military locating utility. The DOD's investment in GPS is immense. Billions and billions of dollars have been invested in creating this technology for military uses. However, over the past several years, GPS has proven to be a useful tool in non-military mapping applications as well.

GPS satellites are orbited high enough to avoid the problems associated with land based systems, yet can provide accurate positioning 24 hours a day, anywhere in the world. Uncorrected positions determined from GPS satellite signals produce accuracies in the range of 50 to 100 meters. When using a technique called differential correction, users can get positions accurate to within five meters or less. Today, many industries are leveraging off the DOD's massive undertaking. As GPS units are becoming smaller and less expensive, there are an expanding number of applications for GPS. In transportation applications, GPS assists pilots and drivers in pinpointing their locations and avoiding collisions. Farmers can use GPS to guide equipment and control accurate distribution of fertilizers and other chemicals. Recreationally, GPS is used for providing accurate locations and as a navigation tool for hikers, hunters and boaters.

Many would argue that GPS has found its greatest utility in the field of Geographic Information Systems (GIS). With some consideration for error, GPS can provide any point on Earth with a unique address i.e. its precise location. A GIS is basically a descriptive database of the Earth or a specific part of the Earth. GPS tells you that you are at point X, Y, Z while GIS tells you that X, Y, Z is a neam tree, or a spot in a stream with a pH level of 5.4. GPS tells us the "where". GIS tells us the "what". GPS/GIS are reshaping the way we locate, organize, analyze and map our resources.

1.1 Determination of Location

GPS is based on satellite ranging - calculating the distances between the receiver and the position of three or more satellites - four or more if elevation is desired - and then applying some good old mathematics. Assuming the positions of the satellites are known, the location of the receiver can be calculated by determining the distance from each of the satellites to the receiver. GPS takes these three or more known references and measured distances and "triangulates" an additional position. As an example, assume that someone is asked you to find him at a stationary position based upon a few clues which he is willing to give you. First, he tells you that he is exactly 10 km away from your house. You would know he is somewhere on the perimeter of a sphere that has an origin as your house and a radius of 10 km. With this information alone, you would have a difficult time to find him since there are an infinite number of locations on the perimeter of that sphere. Second, he tells you that he is also exactly 12 km away from the ABC Grocery Store. Now you can define a second sphere with its origin at the store and a radius of 12 km. You know that he is located somewhere in the space where the perimeters of these two spheres intersect - but there are still many possibilities to define his location.

Adding additional spheres will further reduce the number of possible locations. In fact, a third origin and distance – he tells you are eight km away from the City Clock narrows his position down to just two points. By adding one more sphere, you can pinpoint his exact location. Actually, the fourth sphere may not be necessary. One of the possibilities may not make sense, and therefore can be eliminated. For example, if you know he is above sea level, you can reject a point that has negative elevation. Mathematics and computers allow us to determine the correct point with only three satellites. Based on this example, you can see that you need to know the following information in order to compute your position: A) what is the precise location of three or more known points (GPS satellites)? B) What is the distance between the known points and the position of the GPS receiver?

GPS satellites are orbiting the Earth at an altitude of 20,200 km. The DOD can predict the paths of the satellites vs. time with great accuracy. Furthermore, the satellites can be periodically adjusted by huge land-based radar systems. Therefore, the orbits, and thus the locations of the satellites, are known in advance. Today's GPS receivers store this orbit information for all of the GPS satellites in what is known as an almanac. Think of the almanac as a "bus schedule" advising you of where each satellite will be at a particular time. Each GPS satellite continually broadcasts the almanac. Your GPS receiver will automatically collect this information and store it for future reference.

The Department of Defense, United States of America constantly monitors the orbit of the satellites looking for deviations from predicted values. Any deviations caused by natural atmospheric phenomenon such as gravity, are known as ephemeris errors. When ephemeris errors are determined to exist for a satellite, the errors are sent back up to that satellite, which in turn broadcasts the errors as part of the standard message, supplying this information to the GPS receivers. By using the information from the almanac in conjunction with the ephemeris error data, the position of a GPS satellite can be very precisely determined for a given time.

1.2 Distance between your Position and GPS Satellites

GPS determines distance between a GPS satellite and a GPS receiver by measuring the amount of time it takes a radio signal i.e. the GPS signal to travel from the satellite to the receiver. Radio waves travel at the speed of light, which is about 299,792 km per second. So, if the amount of time it takes for the signal to travel from the satellite to the receiver is known, the distance from the satellite to the receiver i.e. distance = speed x time - can be determined. If the exact time when the signal was transmitted and the exact time when it was received are known, the signal's travel time can be determined.

In order to do this, the satellites and the receivers use very accurate clocks which are synchronized so that they generate the same code at exactly the same time. The code received from the satellite can be compared with the code generated by the receiver. By comparing the codes, the time difference between when the satellite generated the code and when the receiver generated the code can be determined. This interval is the travel time of the code. Multiplying this travel time, in seconds, by 299,792 km per second gives the distance from the receiver position to the satellite.

1.3 Satellites and 3D Position

In the previous example, you saw that it took only three measurements to "triangulate" a 3D position. However, GPS needs a fourth satellite to provide a 3D position. Why? Three measurements can be used to locate a point, assuming the GPS receiver and satellite clocks are precisely and continually synchronized, thereby allowing the distance calculations to be accurately determined. Unfortunately, it is impossible to synchronize these two clocks, since the clocks in GPS receivers are not as accurate as the very precise and expensive atomic clocks in the satellites. The GPS signals travel from the satellite to the receiver very fast, so if the two clocks are off by only a small fraction, the determined position data may be considerably distorted.

The atomic clocks aboard the satellites maintain their time to a very high degree of accuracy. However, there will always be a slight variation in clock rates from satellite to satellite. Close monitoring of the clock of each satellite from the ground permits the control station to insert a message in the signal of each satellite which precisely describes the drift rate of that satellite's clock. The insertion of the drift rate effectively synchronizes all of the GPS satellite clocks.

The same procedure cannot be applied to the clock in a GPS receiver. Therefore, a fourth variable in addition to x, y and z, time, must be determined in order to calculate a precise location. Mathematically, to solve for four unknowns x, y, z, and t, there must be four equations. In determining GPS positions, the four equations are represented by signals from four different satellites.

The US military initially developed GPS to help guide their submarines and missiles, and the navigation satellites are now operated by the US Air Force. To safeguard US national security, the satellites transmit two sets of signals - one for military use and one for civilian use. The commercial market for GPS services has expanded rapidly, with an increasing array of affordable receivers now available for individual use.

1.4 GPS Applications

It didn't take long after the establishment of GPS for the commercial applications to follow. Shipping companies equip their tankers and freighters with GPS for navigation and to record and control the movement of their vessels. Trucking and transportation services use GPS to keep track of their fleets and to speed deliveries. Airlines have saved millions of dollars by using GPS to hone their flight plans. It is used to keep track of government buses and trains in Genoa and Helsinki, and India is examining the possibility of equipping vehicles with GPS receivers to determine when a vehicle is on a road. A dashboard unit would calculate fees and send the information to the toll road operator.

GPS is being used in major cities to manage and track taxis. In some of these taxis, the GPS receivers are linked to an emergency response system which allows the driver to contact police with a precise location. Engineers use GPS for surveying roads and planning buildings and bridges, and it can be used to map natural resources such as soil and vegetation. For example, farmers can use GPS, along with other technologies, to pinpoint low-yielding areas within a paddock and apply corrective treatments.

In 1995, Australia played a leading role in the milestone event of the first certification of a new GPS navigation system for large aircraft. This work was done with a Qantas Boeing 747-400 aircraft, together with special equipment from Air services Australia, and led the way for widespread international use of GPS for many aircraft. So, every time you fly across the Pacific, your aircraft is being guided by GPS. GPS receivers for aviation have inbuilt circuitry to detect if there is a faulty signal from satellites. Indeed, the receiver can even decide which satellite is faulty, exclude it from determining position, and thus allow the flight to proceed using the remaining 'healthy' satellites.

1.5 Future of GPS

It's easy to see that GPS is quickly becoming indispensable to our everyday lives. Consequently, the technology needs to be robust and ultra reliable. In many ways it is, however, does come with some weaknesses. For example, like all radio navigation systems, the signals coming from the satellites are vulnerable to interference. Research into these areas of weakness is leading to improved performance of GPS receivers. Once the use of GPS in cars and mobile phones becomes widespread, 'location services' will be able to offer roadside assistance, traffic updates, and route planning and shopping guides. How to protect the information gathered on your habits and whereabouts is a major privacy issue that will have to be addressed. So the next time you hear a discussion on GPS, don't just pass it off as an interesting bit of technological trivia about satellites many thousands of kilometres away. It's actually a technology that will affect us all.

Chapter - 2

HISTORY OF DEVELOPMENT OF GPS

2.1 Early Methods of Navigation

The shape and size of the Earth has been known from the time of antiquity. The fact that the Earth is a sphere was well known to educated people as long ago as the fourth century BC. In his book On the Heavens, Aristotle gave two scientifically correct arguments. First, the shadow of the Earth projected on the moon during a lunar eclipse appears to be curved. Second, the elevations of stars change as one travels north or south, while certain stars visible in Egypt cannot be seen at all from Greece.

The actual radius of the Earth was determined within one per cent by Eratosthenes in about 230 BC. He knew that the sun was directly overhead at noon on the summer solstice in Syene (Aswan, Egypt), since on that day it illuminated the water of a deep well. At the same time, he measured the length of the shadow cast by a column on the grounds of the library at Alexandria, which was nearly due north. The distance between Alexandria and Syene had been well established by professional runners and camel caravans. Thus Eratosthenes was able to compute the Earth's radius from the difference in latitude that he inferred from his measurement. In terms of modern units of length, he arrived at the figure of about 6400 km. By comparison, the actual mean radius is 6371 km. The Earth is not precisely spherical, as the polar radius is 21 km less than the equatorial radius of 6378 km.

The ability to determine one's position on the Earth was the next major problem to be addressed. In the second century AD the Greek astronomer Claudius Ptolemy prepared a geographical atlas, in which he estimated the latitude and longitude of principal cities of the Mediterranean world. Ptolemy is most famous, however, for his geocentric theory of planetary motion, which was the basis for astronomical catalogs until Nicholas Copernicus published his heliocentric theory in 1543.

Historically, methods of navigation over the Earth's surface have involved the angular measurement of star positions to determine latitude. The latitude of one's position is equal to the elevation of the pole star. The position of the pole star on the celestial sphere is only temporary, however, due to precession of the Earth's axis of rotation through a circle of radius 23.5 over a period of 26,000 years. At the time of Julius Caesar, there was no star sufficiently close to the north celestial pole to be called a pole star. In 13,000 years, the star Vega will be near the pole. It is perhaps not a coincidence that mariners did not venture far from visible land until the era of Christopher Columbus, when true north could be determined using the star we now call Polaris. Even then the star's diurnal rotation caused an apparent variation of the compass needle. Polaris in 1492 described a radius of about 3.5 about the celestial pole, compared to 1 today. At sea, however, Columbus and his contemporaries depended primarily on the mariner's compass and dead reckoning.

The determination of longitude was much more difficult. Longitude is obtained astronomically from the difference between the observed time of a celestial event, such as an eclipse, and the corresponding time tabulated for a reference location. For each hour of difference in time, the difference in longitude is 15 degrees.

Columbus himself attempted to estimate his longitude on his fourth voyage to the New World by observing the time of a lunar eclipse as seen from the harbor of Santa Gloria in Jamaica on February 29, 1504. In his distinguished biography Admiral of the Ocean Sea, Samuel Eliot Morrison states that Columbus measured the duration of the eclipse with an hour-glass and determined his position as seven hours and fifteen minutes west of Cadiz, Spain, according to the predicted eclipse time in an almanac he carried aboard his ship. Over the preceding year, while his ship was marooned in the harbor, Columbus had determined the latitude of Santa Gloria by numerous observations of the pole star. He made out his latitude to be 18, which was in error by less than half a degree and was one of the best recorded observations of latitude in the early sixteenth century, but his estimated longitude was off by some 38 degrees.

Columbus also made legendary use of this eclipse by threatening the natives with the disfavor of God, as indicated by a portent from Heaven, if they did not bring desperately needed provisions to his men. When the eclipse arrived as predicted, the natives pleaded for the Admiral's intervention, promising to furnish all the food that was needed.

New knowledge of the universe was revealed by Galileo in his book The Starry Messenger. This book, published in Venice in 1610, reported the telescopic discoveries of hundreds of new stars, the craters on the moon, the phases of Venus, the rings of Saturn, sunspots, and the four inner satellites of Jupiter. Galileo suggested using the eclipses of Jupiter's satellites as a celestial clock for the practical determination of longitude, but the calculation of an accurate ephemeris and the difficulty of observing the satellites from the deck of a rolling ship prevented use of this method at sea. Nevertheless, James Bradley, the third Astronomer Royal of England, successfully applied the technique in 1726 to determine the longitudes of Lisbon and New York with considerable accuracy.

Inability to measure longitude at sea had the potential of catastrophic consequences for sailing vessels exploring the new world, carrying cargo, and conquering new territories. Shipwrecks were common. On October 22, 1707 a fleet of twenty-one ships under the command of Admiral

Sir Clowdisley Shovell was returning to England after an unsuccessful military attack on Toulon in the Mediterranean. As the fleet approached the English Channel in dense fog, the flagship and three others foundered on the coastal rocks and nearly two thousand men perished.

Stunned by this unprecedented loss, the British government in 1714 offered a prize of £20,000 for a method to determine longitude at sea within a half a degree. The scientific establishment believed that the solution would be obtained from observations of the moon. The German cartographer Tobias Mayer, aided by new mathematical methods developed by Leonard Euler, offered improved tables of the moon in 1757. The recorded position of the moon at a given time as seen from a reference meridian could be compared with its position at the local time to determine the angular position west or east.

Just as the astronomical method appeared to achieve realization, the British craftsman John Harrison provided a different solution through his invention of the marine chronometer. The story of Harrison's clock has been recounted in Dava Sobel's popular book, Longitude. Both methods were tested by sea trials. The lunar tables permitted the determination of longitude within four minutes of arc, but with Harrison's chronometer the precision was only one minute of arc. Ultimately, portions of the prize money were awarded to Mayer's widow, Euler, and Harrison.

In the twentieth century, with the development of radio transmitters, another class of navigation aids was created using terrestrial radio beacons, including Loran and Omega. Finally, the technology of artificial satellites made possible navigation and position determination using line of sight signals involving the measurement of Doppler shift or phase difference.

2.2 Transit

Transit, the Navy Navigation Satellite System, was conceived in the late 1950s and deployed in the mid-1960s.

It was finally retired in 1996 after nearly 33 years of service. The Transit system was developed because of the need to provide accurate navigation data for Polaris missile submarines. As related in an historical perspective by Bradford Parkinson, et al. in the journal Navigation (Spring 1995), the concept was suggested by the predictable but dramatic Doppler frequency shifts from the first Sputnik satellite, launched by the Soviet Union in October, 1957. The Doppler-shifted signals enabled a determination of the orbit using data recorded at one site during a single pass of the satellite. Conversely, if a satellite's orbit were already known, a radio receiver's position could be determined from the same Doppler measurements.

The Transit system was composed of six satellites in nearly circular, polar orbits at an altitude of 1075 km. The period of revolution was 107 minutes. The system employed essentially the same Doppler data used to track the Sputnik satellite. However, the orbits of the Transit satellites were precisely determined by tracking them at widely spaced fixed sites. Under favorable conditions, the rms accuracy was 35 to 100 meters. The main problem with Transit was the large gaps in coverage. Users had to interpolate their positions between passes.

2.3 Global Positioning System

The success of Transit stimulated both the U.S. Navy and the U.S. Air Force to investigate more advanced versions of a space-based navigation system with enhanced capabilities. Recognizing the need for a combined effort, the Deputy Secretary of Defense established a Joint Program Office in 1973. The NAVSTAR Global Positioning System (GPS) was thus created. In contrast to Transit, GPS provides continuous coverage. Also, rather than Doppler shift, satellite range is determined from phase difference. There are two types of observables. One is pseudo range, which is the offset between a pseudorandom noise (PRN) coded signal from the satellite and a replica code generated in the user's receiver, multiplied by the speed of light. The other is accumulated delta range (ADR), which is a measure of carrier phase.

The determination of position may be described as the process of triangulation using the measured range between the user and four or more satellites. The ranges are inferred from the time of propagation of the satellite signals. Four satellites are required to determine the three coordinates of position and time. The time is involved in the correction to the receiver clock and is ultimately eliminated from the measurement of position.

High precision is made possible through the use of atomic clocks carried on-board the satellites. Each satellite has two cesium clocks and two rubidium clocks, which maintain time with a precision of a few parts in 1013 or 1014 over a few hours, or better than 10 nanoseconds. In terms of the distance traversed by an electromagnetic signal at the speed of light, each nanosecond corresponds to about 30 centimeters. Thus the precision of GPS clocks permits a real time measurement of distance to within a few meters. With post-processed carrier phase measurements, a precision of a few centimeters can be achieved.

The design of the GPS constellation had the fundamental requirement that at least four satellites must be visible at all times from any point on Earth. The tradeoffs included visibility, the need to pass over the ground control stations in the United States, cost, and sparing efficiency. The orbital configuration approved in 1973 was a total of 24 satellites, consisting of 8 satellites plus one spare in each of three equally spaced orbital planes. The orbital radius was 26,562 km, corresponding to a period of revolution of 12 sidereal hours, with repeating ground traces. Each satellite arrived over a given point four minutes earlier each day. A common orbital inclination of 63 was selected to maximize the on-orbit payload mass with launches from the Western Test Range. This configuration ensured between 6 and 11 satellites in view at any time.

As envisioned ten years later, the inclination was reduced to 55 and the number of planes was increased to six. The constellation would consist of 18 primary satellites, which represents the absolute minimum number of satellites required to provide continuous global coverage with at least four satellites in view at any point on the Earth. In addition, there would be three on-orbit spares. The operational system, as presently deployed, consists of 21 primary satellites and three on-orbit spares, comprising four satellites in each of six orbital planes. Each orbital plane is inclined at 55. This constellation improves on the "18 plus 3" satellite constellation by more fully integrating the three active spares.

There have been several generations of GPS satellites. The Block I satellites, built by Rockwell International, were launched between 1978 and 1985. They consisted of eleven prototype satellites, including one launch failure that validated the system concept. The ten successful satellites had an average lifetime of 8.76 years.

The Block II and Block IIA satellites were also built by Rockwell International. Block II consists of nine satellites launched between 1989 and 1990. Block IIA, deployed between 1990 and 1997, consists of 19 satellites with several navigation enhancements. In April 1995, GPS was declared fully operational with a constellation of 24 operational spacecraft and a completed ground segment. The 28 Block II/IIA satellites have exceeded their specified mission duration of six years and are expected to have an average lifetime of more than ten years. Block IIR comprises 20 replacement satellites that incorporate autonomous navigation based on crosslink ranging. These satellites are being manufactured by Lockheed Martin. The first launch in 1997 resulted in a launch failure. The first IIR satellite to reach orbit was also launched in 1997. The second GPS IIR satellite was successfully launched aboard a Delta 2 rocket on October 7, 1999.

The fourth generation of satellites is the Block II follow-on (Block IIF). This program includes the procurement of 33 satellites and the operation and support of a new GPS operational control segment. The Block IIF program was awarded to Rockwell (now a part of Boeing). Further details may be found in a special issue of the Proceedings of the IEEE for January, 1999. The Master Control Station for GPS is located at Schriever Air Force Base in Colorado Springs, CO. The MCS maintains the satellite constellation and performs the station keeping and attitude control maneuvers. It also determines the orbit and clock parameters with a Kalman filter using measurements from five monitor stations distributed around the world. The orbit error is about 1.5 meters.

GPS orbits are derived independently by various scientific organizations using carrier phase and post-processing. The state of the art is exemplified by the work of the International GPS Service (IGS), which produces orbits with an accuracy of approximately three centimeters within two weeks. The system time reference is managed by the U.S. Naval Observatory in Washington, DC. GPS time is measured from Saturday/Sunday midnight at the beginning of the week. The GPS time scale is a composite "paper clock" that is synchronized to keep step with Coordinated Universal Time (UTC) and International Atomic Time (TAI). However, UTC differs from TAI by an integral number of leap seconds to maintain correspondence with the rotation of the Earth, whereas GPS time does not include leap seconds. The origin of GPS time is midnight on January 5/6, 1980(UTC). At present, TAI is ahead of UTC by 32 seconds, TAI is ahead of GPS by 19 seconds, and GPS is ahead of UTC by 13 seconds. Only 1,024 weeks were allotted from the origin before the system time is reset to zero because 10 bits are allocated for the calendar function (1,024 is the tenth power of 2). Thus the first GPS rollover occurred at midnight on August 21, 1999. The next GPS rollover will take place May 25, 2019.

2.4 GPS Modernization

In 1996, a Presidential Decision Directive stated the president would review the issue of Selective Availability in 2000 with the objective of discontinuing SA no later than 2006. In addition, both the L1 and L2 GPS signals would be made available to civil users and a new civil 10.23 MHz signal would be authorized. To satisfy the needs of aviation, the third civil frequency, known as L5, would be centered at 1176.45 MHz, in the Aeronautical Radio Navigation Services (ARNS) band, subject to approval at the World Radio Conference in 2000. According to Keith McDonald in an article on GPS modernization published in the September, 1999 GPS World, with SA removed the civil GPS accuracy would be improved to about 10 to 30 meters. With the addition of a second frequency for ionospheric group delay corrections, the civil accuracy would become about 5 to 10 meters. A third frequency would permit the creation of two beat frequencies that would yield one-meter accuracy in real time.

A variety of other enhancements are under consideration, including increased power, the addition of a new military code at the L1 and L2 frequencies, additional ground stations, more frequent uploads, and an increase in the number of satellites. These policy initiatives are driven by the dual needs of maintaining national security while supporting the growing dependence on GPS by commercial industry. When these upgrades would begin to be implemented in the Block IIR and IIF satellites depends on GPS funding.

Besides providing position, GPS is a reference for time with an accuracy of 10 nanoseconds or better. Its broadcast time signals are used for national defense, commercial, and scientific purposes. The precision and universal availability of GPS time has produced a paradigm shift in time measurement and dissemination, with GPS evolving from a secondary source to a fundamental reference in itself.

The international community wants assurance that it can rely on the availability of GPS and continued U.S. support for the system. The Russian Global Navigation Satellite System (GLONASS) has been an alternative, but economic conditions in Russia have threatened its continued viability. Consequently, the European Union is considering the creation of a navigation system of its own, called Galileo, to avoid relying on the U.S. GPS and Russian GLONASS programs.

The Global Positioning System is a vital national resource. Over the past thirty years it has made the transition from concept to reality, representing today an operational system on which the entire world has become dependent. Both technical improvements and an enlightened national policy will be necessary to ensure its continued growth into the twenty-first century.

Chapter - 3

GLOBAL POSITIONING SYSTEM - AN OVERVIEW

The Global Positioning System consists of 24 Earth-orbiting satellites so that it can guarantee that there are at least four of them above the horizon for any point on Earth at any time. In general there are normally eight or so satellites "visible" to a GPS receiver at any given moment. Each satellite contains an atomic clock. The satellites send radio signals to GPS receivers so that the receivers can find out how far away each satellite is — sometimes with millimetres precision — at a given time. Because the satellites are orbiting at a distance of 20,200 km overhead, the signals are fairly weak by the time they reach a GPS receiver. That means we have to be outside in a fairly open area for the GPS receiver to work.

Receiving GPS devices range from hand-held units to more sophisticated vehicle-mounted and stationary equipment. Trilateration or resection is a basic geometric principle that allows us to find one location if we know its distance from other, already known locations. The heart of a GPS receiver is the ability to find the receiver's distance from four or more GPS satellites. Once it determines its distance from the four satellites, the receiver can calculate its exact location and altitude on Earth. If the receiver can only find three satellites, then it can use an imaginary sphere to represent the Earth and can give you location information but no altitude information. GPS satellites send out radio signals that a GPS receiver can detect. For the GPS receiver to know how far away the satellite is, it measures the amount of time it takes for the signal to travel from the satellite to the receiver. Since we know how fast radio signals travel – they are electromagnetic waves traveling at the speed of light. We can figure out how far they have traveled by figuring out how long it took for them to arrive.

Measuring the time would be easy if we knew exactly what time the signal left the satellite and exactly what time it arrived at the receiver. Solving this problem is key to the Global Positioning System. One way to solve the problem would be to put extremely accurate and synchronized clocks in the satellites and the receivers. The satellite begins transmitting a long digital pattern, called a Pseudo Random Code, as part of its signal at a certain time, let's say midnight. The receiver begins running the same digital pattern, also exactly at midnight. When the satellite's signal reaches the receiver, its transmission of the pattern will lag a bit behind the receiver's playing of the pattern. The length of the delay is equal to the time of the signal's travel. The receiver multiplies this time by the speed of light to determine how far the signal traveled. If the signal traveled in a straight line, this distance would be the distance to the satellite.

The only way to implement a system like this would require a level of accuracy only found in atomic clocks. This is because the time measured in these calculations amounts to nanoseconds. To make a GPS using only synchronized clocks, we would need to have atomic clocks not only on all the satellites, but also in the receiver itself. Atomic clocks usually cost somewhere between £30,000 and £60,000, which makes them too expensive for everyday consumer use.

The Global Positioning System has a very effective solution to this problem – a GPS receiver contains no atomic clock at all. It has a normal quartz clock. The receiver looks at all the signals it is receiving and uses calculations to find both the exact time and the exact location simultaneously. When we measure the distance to four located satellites, we can draw four spheres that all intersect at one point, as illustrated above. Four spheres of this sort will not intersect at one point if we have measured incorrectly. Since the receiver makes all of its time measurements, and therefore its distance measurements, using the clock it is equipped with, the distances will all be proportionally incorrect. The receiver can therefore easily calculate exactly what distance adjustment will cause the four spheres to intersect at one point. This allows it to adjust its clock to adjust its measure of distance. For this reason, a GPS receiver actually keeps extremely accurate time, on the order of the actual atomic clocks in the satellites.

One problem with this method is the measure of speed. As we saw earlier, electromagnetic signals travel through a vacuum at the speed of light. The Earth, of course, is not a vacuum, and its atmosphere slows the transmission of the signal according to the particular conditions at that atmospheric location, the angle at which the signal enters it, and so on. A GPS receiver guesses the actual speed of the signal using complex mathematical models of a wide range of atmospheric conditions. The satellites can also transmit additional information to the receiver.

The other crucial component of GPS calculations is the knowledge of where the satellites are. This isn't difficult because the satellites travel in a very high and predictable orbit. The satellites are far enough from the Earth that they are not affected by our atmosphere. The GPS receiver simply stores an almanac that tells it where every satellite should be at any given time. Things like the pull of the moon and the sun do change the satellites' orbits very slightly, but the US Department of Defense constantly monitors their exact positions and transmits any adjustments to all GPS receivers as part of the satellites' signals.

The most essential function of a GPS receiver is to pick up the transmissions of at least four satellites and combine the information in those transmissions with information in an electronic almanac, so that it can mathematically determine the receiver's position on Earth. The basic information a receiver provides, then, is the latitude, longitude and altitude or some similar measurement of its current position. Most receivers then combine this data with other information, such as maps, to make the receiver more user-friendly. We can use maps stored in the receiver's memory, connect the receiver up to a computer that can hold more detailed maps in its memory or simply buy a detailed map of our area and find our way using the receiver's latitude and longitude readouts.

3.1 System Overview

The total GPS configuration is comprised of three distinct segments:

The Space Segment	-	Satellites orbiting the Earth.
The Control Segment	-	Stations positioned on the Earth's Equator to control the satellites
The User Segment	-	Any body that receives and uses the GPS signal.

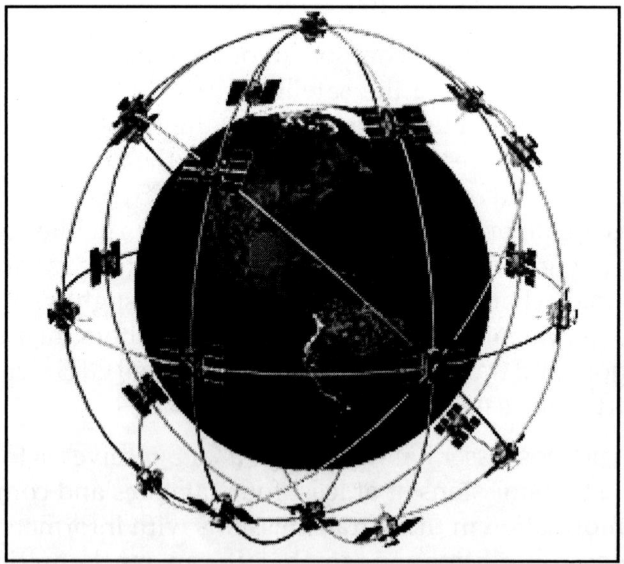

Fig. 3.1: GPS Satellite Constellation

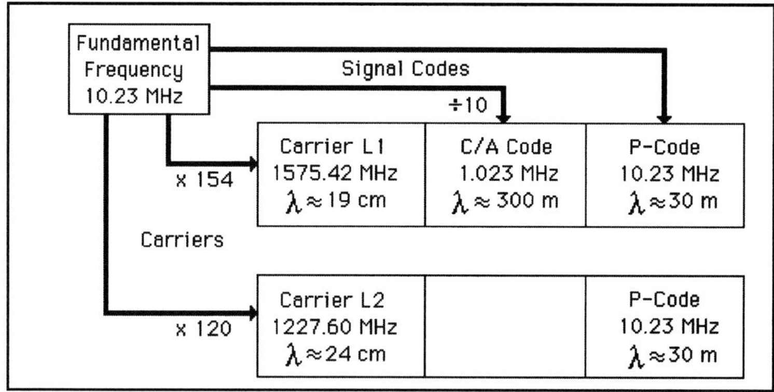

Fig. 3.2: GPS Signal Structure

Fig.3.3: Network of GPS Master Control and Monitoring Stations

3.2 Space Segment

The Space Segment of the system consists of the GPS satellites. These space vehicles (SVs) send radio signals from space. The nominal GPS Operational Constellation consists of 24 satellites orbiting the Earth at approximately 20200 km every 12 hours (Fig.3.1). There are often more than 24 operational satellites as new ones are launched to replace older satellites. The satellite orbits repeat almost the same ground track as the Earth turns beneath them once each day. The orbit altitude is such that the satellites repeat the same track and configuration over any point approximately each 24 hours four minutes earlier each day. There are six orbital planes with nominally four SVs in each, equally spaced 60 degrees apart, and inclined at about fifty-five degrees with respect to the equatorial plane. This constellation provides the user with between five and eight SVs visible from any point on the Earth.

Each GPS satellite has several very accurate atomic clocks on board. The clocks operate at a fundamental frequency of 10.23 MHz. This is used to generate the signals (Fig. 3.2) that are broadcast from the satellite. The satellites broadcast two carrier waves constantly. These carrier waves are in the L-Band used for radio, and travel to Earth at the speed of light. These carrier waves are derived from the fundamental frequency, generated by a very precise atomic clock:

The L_1 carrier is broadcast at 1575.42 MHz (10.23 x 154)

The L_2 carrier is broadcast at 1227.60 MHz (10.23 x 120).

The L1 carrier then has two codes modulated upon it. The C/A Code or Coarse/Acquisition Code is modulated at 1.023MHz (10.23/10) and the P-code or Precision Code is modulated at 10.23MHz. The L2 carrier has just one code modulated upon it. The L2 P-code is modulated at 10.23 MHz. GPS receivers use the different codes to distinguish between satellites. The codes can also be used as a basis for making pseudo range measurements and therefore calculate a position.

3.3 Control Segment

The Control Segment consists of a system of tracking stations located around the world. The Control Segment consists of one master control station, five monitor stations and four ground antennas distributed amongst five locations roughly on the Earth's equator. The Control Segment tracks the GPS satellites, updates their orbiting position and calibrates and synchronizes their clocks. A further important function is to determine the orbit of each satellite and predict its path for the following 24 hours. This information is uploaded to each satellite and subsequently broadcast from it. This enables the GPS receiver to know where each satellite can be expected to be found. The satellite signals are read at Ascension, Diego Garcia and Kwajalein. The measurements are then sent to the Master Control Station in Colorado Springs where they are processed to determine any errors in each satellite. The information is then sent back to the four monitor stations equipped with ground antennas and uploaded to the satellites.

3.4 User Segment

The User Segment comprises of anyone using a GPS receiver to receive the GPS signal and determine their position and/or time. Typical applications within the user segment are land navigation for hikers, vehicle location, surveying, marine navigation, aerial navigation, machine control etc.

3.5 Navigation and Positioning Advantages

1. High positioning accuracies.
2. Determination of velocity and time to accuracy commensurate with position.
3. Available to users anywhere on the globe: in the air, on the ground, or at sea.
4. Relatively low cost system, with no user charges.
5. All-weather system, available 24 hours a day. Can transmit signals through cloud and rain.
6. The position information is in three dimensions, that is, vertical(Altitude) as well as horizontal (Latitude and Longitude) information is provided
7. Can transmit signals that can cover a far larger area than ground-based systems.
8. Can be used day or night, as long as the transmitting satellite is above the user's horizon.
9. Recognizes no national boundaries, and refers positions to a global datum uniquely defined for that system.

3.6 Advantages of GPS over Conventional Surveying Methods

There are several advantages of the GPS satellite surveying technique:

1. Intervisibility between stations is not necessary.
2. Because GPS uses radio frequencies to transmit the signals, the system is independent of weather conditions.
3. If the same field and data reduction procedures are used, position accuracy is largely a function of interstation distance, and not of network "shape" or "geometry".
4. Because of the generally homogeneous accuracy of GPS surveying, geodetic network planning in the classical sense is no longer relevant. The points are placed where they are required for example, in a valley, and need not be located at evenly distributed sites atop mountains to satisfy intervisibility, or network geometry, criteria.
5. Because of the two advantages of not requiring intervisibility of stations, or following a conventional network design strategy, GPS surveying is more efficient, more flexible and less time consuming a positioning technique than using terrestrial survey technologies.
6. GPS can be used around-the-clock.
7. GPS provides three-dimensional information.
8. High accuracies can be achieved with relatively little effort, unlike conventional terrestrial techniques. The GPS instrumentation, and to some extent the data processing software, is similar whether accuracies at the 1 part in 10^4 or 1 part in 10^6 level are sought.

3.7 Disadvantages of GPS Surveying

It would be remiss not to also mention the disadvantages, some of which will no doubt be overcome in time, others with some additional effort, while others cannot be dismissed so easily:

1. High efficiency has its price. Efficient use of GPS requires that travel times between stations are cut in order to match the savings in on-site time.
2. Because station intervisibility is not necessary GPS is a particularly attractive technology for use in rugged, inhospitable terrain. However, the logistical problems of transporting and supporting several field parties are still formidable and would have been even if conventional terrestrial techniques were used. If helicopters are necessary, the costs of the survey will rise substantially.
3. GPS requires that there be no obstruction to the signals by overhanging branches or structures though the antenna can be raised above the obstruction. It cannot, of course, be used underground, and may have limited application in densely settled urban areas.
4. Because GPS surveys can be "optimized" by appropriate selection of sites to satisfy the specific needs of the particular survey, these may not be useful for other applications in the same area. Further GPS surveys may need to be carried out, as the need arises, for new applications. The extreme of this would be to dispense with a permanently monumented control network altogether, and to require the re-establishment of coordinates by GPS each time the need arises.
5. Two intervisible stations would have to establish by GPS in order to satisfy the requirement for azimuth data for use by conventional line-of-sight survey methods.
6. GPS coordinates are provided in the Earth-centred, Earth-fixed coordinate system defined by the GPS

satellite ephemeredes - the WGS84 system when the broadcast ephemeris is used. Results may need to be transformed into a local geodetic system before they can be integrated with results from conventional surveys.

7. GPS results are, in general, more accurate than the surrounding control marks established by terrestrial techniques over time. Comparison of GPS and terrestrial results will be the source of confusion, controversy and conflict for many years to come.
8. GPS vertical information is not provided in the height system generally required. The GPS heights have to be reduced to a sea level datum more precisely, the geoid.
9. The GPS instrumentation is still comparatively expensive. Although the price of one receiver is likely to soon match that of a theodolite-EDM instrument, a minimum of two are required for survey work.
10. GPS requires new skills to be learned, and new procedures and strategies for planning, field operation and data analysis to be developed. In addition, an understanding of how GPS results can be integrated with conventional horizontal and vertical networks is required.

Geographers have mapped every corner of the Earth, so we can certainly find maps with the desired level of detail. We can look at a GPS receiver as an extremely accurate way to get raw positional data, which can then be applied to geographic information that has been accumulated over the years. We can even use GPS to accurately update older maps and record new/changed features on them.

These systems will change our lives in many ways. By combining GPS with current and future computer mapping techniques, we will be better able to identify and manage our natural resources. Intelligent vehicle location and navigation systems will let us avoid congested freeways and find more efficient routes to our destinations, saving millions of rupees spent in fuel and tons of air pollution. Travel abroad ships and aircraft will be safer in all weather conditions. Businesses with large amounts of outside plant, railroads, utilities will be able to manage their resources more efficiently, reducing consumer costs.

Chapter - 4

BASIC PRINCIPLES OF GPS

This chapter explains various principles which are the basics in understanding the Principles and applications of Global Positioning System. Which includes trilateration, four unknowns, calculating the distance to the satellites, calculating time, height measurements, code and carrier phase GPS, Errors, accuracy and Precision, Techniques to improve accuracy, satellite geometry, various software and hardware needs of GPS.

4.1 Four Unknowns

The problem with GPS is that only pseudo ranges and the time at which the signal arrived at the receiver can be determined. Thus there are four unknowns to determine; position (X, Y, Z) and time of travel of the signal. Observing to four satellites produces four equations which can be solved, enabling these unknowns to be determined.

4.2 Trilateration

All GPS positions are based on measuring the distance from the satellites to the GPS receiver on the Earth. This distance to each satellite can be determined by the GPS receiver. If you know the distance to three points relative to your own position, you can determine your own position relative to those three points. From the distance to one satellite we know that the position of the receiver must be at some point on the surface of an imaginary sphere which

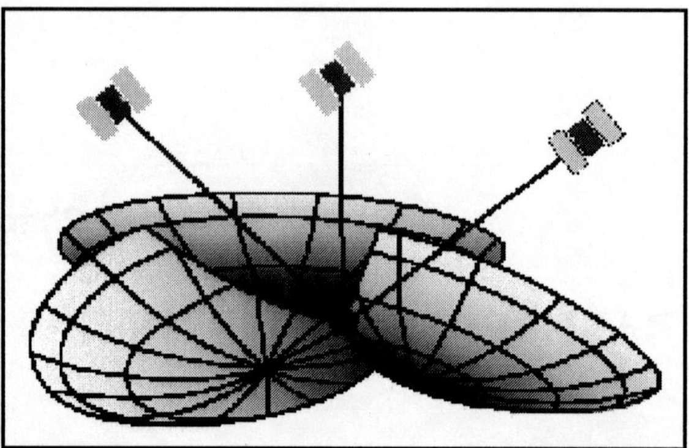

Fig. 4.1: Intersection of Three Imaginary Spheres

has it's origin at the satellite. By intersecting three imaginary spheres the receiver position can be determined.

4.3 Calculating the Distance to the Satellite

In order to calculate the distance to each satellite, one of Isaac Newton's laws of motion is used:

Distance = Velocity X Time

GPS requires the receiver to calculate the distance from the receiver to the satellite. The velocity is the velocity of the radio signal. Radio waves travel at the speed of light, 3×10^8 m/sec.

The Time is the time taken for the radio signal to travel from the satellite to the GPS receiver. This is a little harder to calculate, since you need to know when the radio signal left the satellite and when it reached the receiver.

The satellite signal has two codes modulated upon it; the C/A code and the P-code. The C/A code is based upon the time given by a very accurate atomic clock. The receiver also contains a clock that is used to generate a matching C/A code. The GPS receiver is then able to "match" or correlate the incoming satellite code to the receiver generated code.

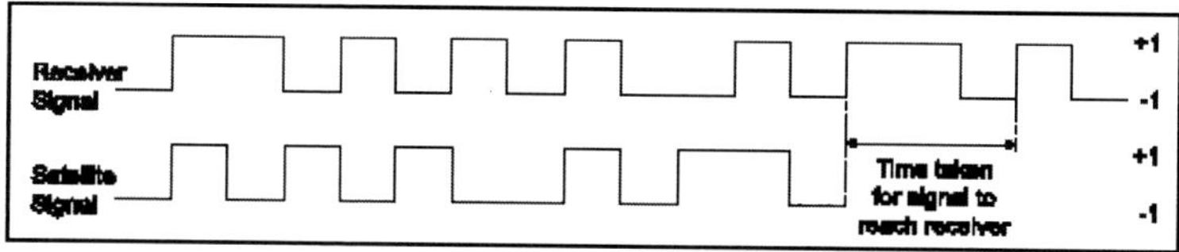

Fig. 4.2: Calculating Time

The C/A code is a digital code that is "pseudo random" or appears to be random. In actual fact it is not random and repeats one thousand times every second. In this way, the time taken for the radio signal to travel from the satellite to the GPS receiver is calculated.

4.4 Height Measurements

Heights are traditionally determined using leveling techniques which are based on the Earth's gravity field and referenced to mean sea level. At each point on the Earth, gravity has a certain magnitude and direction which can be described by a vector. Every time an instrument is leveled, the line of sight is set perpendicular to the gravity vector at that point, and every time a leveling rod is held on a point, it is held in line with the gravity vector. The heights determined through leveling are usually referred to as elevations above mean sea level. These elevations are called

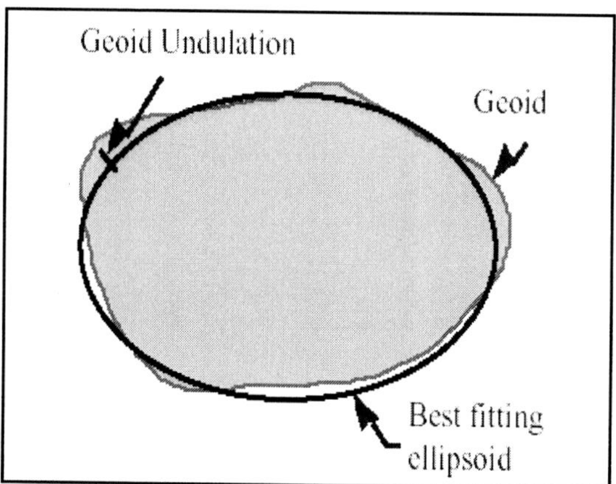

Fig. 4.3: Relationship between the Ellipsoid and Geoid

orthometric heights and actually are referenced to the geoid. Orthometric heights are the heights (or elevations) which see everyday use and are found on topographical maps.

The geoid is the equipotential surface (i.e. the surface on which the gravity potential is constant) which best approximates mean sea level. It forms a smooth but irregular surface around the Earth. Mathematically, it is a very complex surface to represent. On the other hand, the ellipsoid, which is in essence a squashed sphere, is easily represented and manipulated mathematically. It is for this reason that an ellipsoid is used to approximate the geoid, and that ellipsoidal heights based on the ellipsoid are determined rather than geoid heights based on the geoid. Elevations determined using GPS satellites are based on the ellipsoidal surface. The relationship between the ellipsoid and geoid is illustrated in Figure 4.3.

The distance separating the geoid and the ellipsoid is the geoid undulation (also called geoid height). The geoid undulation may be positive or negative depending on whether the geoid is above or below the ellipsoid at a given point. If the geoid undulation, N, and the ellipsoidal height, h, are known, the orthometric height may be determined using the relationship illustrated in the Figure 4.4. That is,

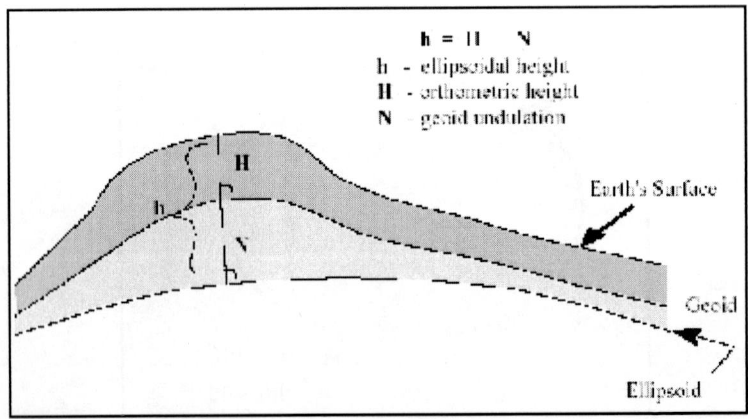

Fig. 4.4: Calculating Ellipsoidal Height

the orthometric height is equal to the ellipsoidal height minus the geoid undulation.

From this it is evident that one must know geoid undulations to compute orthometric heights with GPS. Geoid undulations vary by as much as ±100 m over the Earth's surface. Geoid models, which describe the pattern of geoid undulations over the Earth, provide the link between orthometric and ellipsoidal heights. Numerous geoid models have been produced by geodesists by combining scientific theory and various types of gravity measurements. With a geoid model, given a specific latitude and longitude, one can look up or interpolate the geoid undulation.

Geoid models have varying levels of accuracy and areas of coverage. In general, the more accurate the geoid model, the more computationally intensive it is to produce, and the more storage space required. Some geoid models are available for the whole world, and some only for specific regions. For example, the Geodetic Survey Division of EMR has computed a geoid, GSD91, which covers only Canada, but has a level of accuracy and detail which exceeds any available global geoid models and is one of the most accurate models available for Canada. It is available from the Geodetic Survey Division (GSD). GSD also provides geoid information for other parts of the world.

Many GPS receivers, GPS software, or both have "built-in" geoid models. That is, they have the means to correct for the geoid - ellipsoid separation at any given position determined with GPS. These models are usually low accuracy models which apply to the full Earth's surface, and have rms error of about 1.0 m. With low accuracy hand held GPS receivers, the conversion from ellipsoid to orthometric heights may even be made transparent to the user. The need for and accuracy level required of geoid models depend on whether single point or relative positioning is used. For single point positioning a geoid model must always be used if orthometric heights are sought. However, since single point positioning is only accurate to 100 m horizontally and 156 m vertically the effect of geoid undulation inaccuracies at the 1.0 m rms level is negligible.

For relative positioning, the situation is very different. Relative positioning always involves determining an unknown point relative to a known point. Hence it is not only the absolute geoid undulation that is of concern, but also the difference between the geoid undulations at the known and unknown points. The use of a geoid model is more complex and variable depending of the accuracy level required. The ellipsoid is also the basis for all horizontal measurements made using GPS. This brings us to discussion of coordinate systems and datums as they relate to GPS.

4.5 Calculating Time

GPS has become the world's principal supplier of accurate time. It is used extensively both as a source of time and as a means of transferring time from one location to another. There are three kinds of time available from GPS: GPS time, UTC as estimated and produced by the United States Naval Observatory, and the times from each free-running GPS satellite's atomic clock. The Master Control Station (MCS) at Falcon Air Force Base near Colorado Springs, Colorado gathers the GPS satellites' data from five monitor stations around the globe. A Kalman filter software program estimates the time error, frequency error, frequency drift and Keplerian orbit parameters for each of the satellites and its operating clock. This information is uploaded to each satellite so that it can be broadcasted in real time. This process provides GPS time consistency across the constellation to within a small number of nanoseconds and accurate position determination of the satellites to within a few meters. In order to provide an estimate of UTC time derivable from a GPS signal, a set of UTC corrections is also provided as part of the broadcast signal. This broadcast message includes the time difference in whole seconds between GPS time and UTC. During 1996 GPS time minus UTC time was 11 seconds.

GPS time is a standard time scale used to correlate the atomic clocks on the satellites to the clock in the receiver unit. GPS time is referenced to the master clock (MC) at the U.S. Navel Observatory (USNO) and does not deviate by more than 1 microsecond from the coordinated universal time (UTC) worldwide time scale. USNO monitors the timing of the GPS to provide a reliable and stable coordinated time reference for the satellite navigation system. The USNO provides two modes of operation to monitor the GPS: the Standard Positioning Service (SPS) and the Precise Positioning Service (PPS). The USNO SPS consists of a coarse acquisition (C/A) code, single channel timing receiver and the processed data are available on the USNO timing data web site http://tycho.usno.navy.mil/gps_datafiles.html.

Each Block II/IIA satellite contains two cesium (Cs) and two rubidium (Rb) atomic clocks. Each Block IIR satellite contains three Rb atomic clocks. GPS time is given by its Composite Clock (CC). The CC or "paper" clock consists of all Monitor Station and satellite operational frequency standards. The system was previously referenced to one of the Monitor Station's operational frequency standards and switched from one station to another as needed. The GPS epoch is 0000 UT (midnight) on January 6, 1980. GPS time is not adjusted and therefore is offset from UTC (Coordinated Universal Time) by an integer number of seconds, due to the insertion of leap seconds (Appendix : II). The number remains constant until the next leap second occurs. This offset is also given in the navigation (NAV) message and the receiver should apply the correction automatically. As of January 1, 1999, GPS time is ahead of UTC by thirteen (13) seconds.

In addition to the leap seconds, there are additional corrections given in the NAV message. The system time, in turn, is referenced to the Master Clock (MC) at the USNO and steered to UTC (USNO) from which time will not deviate by more than one microsecond (PPS requirement). The exact difference is contained in the NAV message in the form of two constants, A0 and A1, giving the time difference and rate of system time against UTC (USNO, MC). UTC (USNO) itself is kept very close to the international benchmark UTC. A direct reference to UTC (USNO, MC) can be made automatically by most timing receivers. These receivers can be commanded to take the two constants, A0 and A1, from the NAV message for a linear extrapolation to the USNO MC. These constants are updated with the uploads on the basis of USNO PPS monitor information.

GPS time is automatically steered to UTC (USNO) on a daily basis to keep system time within one microsecond of UTC (USNO), but during the last several years has been within a few hundred nanoseconds. The rate of steer being applied is +/-1.0E-19 seconds per second squared.

All GPS satellites are monitored at USNO, collecting time transfer data for a track period of 780 seconds duration. The individual values are an estimate of the difference between the USNO MC and the GPS CC (GPS time) via the individual satellite. The RMS of the 13-minute solutions ranges between 2 to 20 nanoseconds Block II without Selective Availability (SA). When SA was implemented on the Block II constellation, the RMS ranged from 20 to 100 nanoseconds. Daily overall values for the entire constellation are an estimate of the difference between the USNO MC and the GPS CC. These values represent a 2-day filtered linear solution and computed for zero hours UT of the second day and published daily for the preceding day. The RMS of the residuals for the entire constellation when SA is not implemented is around 4 to 10 nanoseconds. When SA was implemented, the RMS of the residuals ranged from 40 to 60 nanoseconds for the entire constellation.

4.6 Code and Carrier Phase GPS

The words "Code-Phase" and "Carrier-Phase" may sound strange but, in fact, they just refer to the particular signal that is used for timing measurements. Using the GPS carrier frequency can significantly improve the accuracy of GPS. The concept is simple. It is already discussed that a GPS receiver determines the travel time of a signal from a satellite by comparing the "pseudo random code" it generates, with an identical code in the signal from the satellite.

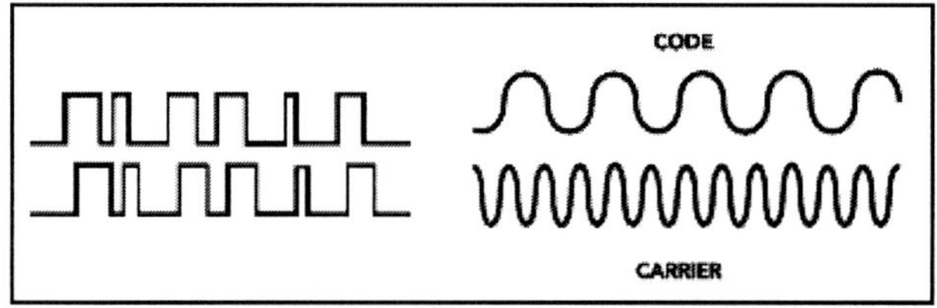

Fig. 4.5: Code and Carrier Phase GPS

The receiver slides its code for some time till it synchronized up with the satellite's code. The amount it has to slide the code is equal to the signal's travel time. The problem is that the bits (or cycles) of the pseudo random code are so wide that even if the receiver synchronized up there's still plenty of slop. This is the problem with code-phase GPS. The comparing pseudo random codes have a cycle width of almost a microsecond. And at the speed of light a microsecond is almost 300 meters of error! However, this doesn't mean that code phase GPS is bad. Because receiver designers have come up with ways to make sure that the signals are almost perfectly in phase. Good machines get with in a percent or two. But that's still at least 3-6 meters of error.

Survey receivers overcome this problem by starting with the pseudo random code and then move on to measurements based on the carrier frequency for that code. This carrier frequency is much higher so its pulses are much closer together and therefore more accurate.

At this point it is possible to make some brief comparisons of code and carrier measurements. Carrier wavelengths (19 cm for L1) are much shorter than the C/A code chip length

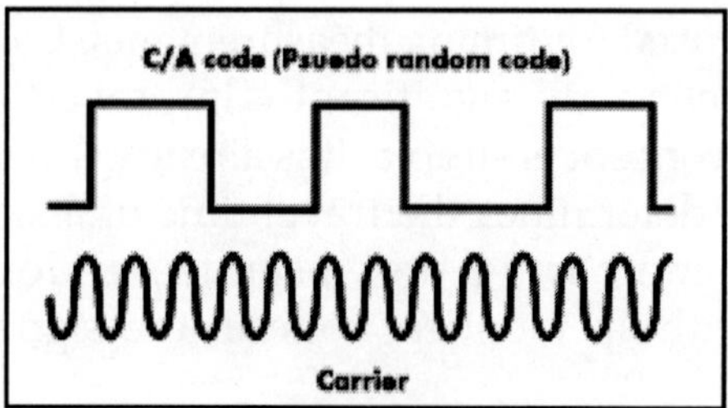

Fig. 4.6: Psuedo Random Code

(293 m) and consequently can be measured more accurately and used to achieve much higher positional accuracies than code measurements. Indeed the best relative accuracies achieved using code measurements are usually a few metres, and using carrier measurement are usually a few centimeters. With code observations a direct measure of the satellite-receiver range is attained. With carrier observations, the ambiguity term (number of whole cycles) must be estimated before one may take advantage of the carrier accuracy. Ambiguity estimation leads to complexities in the use of carrier phase observations which do not exist with code observations. The advantages and disadvantages of code and carrier observations are summarized in the table bellow.

Key Advantages and Disadvantages of Code and Carrier Observations

	Code	**Carrier**
Advantages	non-ambiguous simple	high accuracy potential
Disadvantages	low accuracy	more complex

4.7 Geodetic Aspects

Since GPS has become increasingly popular as a surveying and navigation instrument, surveyors and navigators require a basic understanding of how GPS positions relate to standard mapping systems. A common cause of errors in GPS surveys is the result of incorrectly understanding these relationships.

Determining a position with GPS achieves a fundamental goal of Geodesy - the determination of absolute position with uniform accuracy at all points on the Earth's surface. Using classical geodetic and surveying techniques, determination of position is always relative to the starting points of the survey. The accuracy achieved is dependent on the distance from this point. GPS, therefore, offers a significant advantage over conventional techniques.

The science of Geodesy is basic to GPS, and, conversely, GPS has become a major tool in Geodesy. The following are the aims of geodesy.

1. Establishment and maintenance of national as well as global geodetic control networks on land, recognizing the time-varying nature of these networks due to plate movement.
2. Measurement and representation of geodynamic phenomena-polar motion, Earth tides, and crustal motion.
3. Determination of the gravity field of the Earth including temporal variations.

Although most users would never carry out any of the above tasks, it is essential that users of GPS equipment should have a general understanding of Geodesy.

4.7.1 GPS Coordinate System

Although the Earth may appear to be a uniform sphere when viewed from space, the surface is far from uniform. Due to the fact that GPS has to give coordinates at any point on the Earth's surface, it uses a geodetic coordinate system based on an ellipsoid. An ellipsoid, also known as a spheroid, is a sphere that has been flattened or squashed. An ellipsoid is chosen that most accurately approximates to the shape of the Earth. This ellipsoid has no physical surface but is a mathematically defined surface. There are actually many different ellipsoids or mathematical definitions of the Earth's surface. The ellipsoid used by GPS is known as WGS84 or World Geodetic System 1984. A point on the surface of the Earth can be defined by using latitude, longitude and ellipsoidal height. An alternative method for defining the position of a point is the Cartesian coordinate system, using distances in the X, Y, and Z axes from the origin or centre of the spheroid. This is the method primarily used by GPS for defining the location of a point in space.

4.7.2 World Geodetic System 1984

The datum used for GPS positioning is called WGS84 (World Geodetic System 1984). It consists of three-dimensional Cartesian coordinate system and an associated ellipsoid, so that WGS84 positions can be described as XYZ Cartesian coordinates or latitude, longitude and ellipsoid height coordinates. The origin of the datum is the Geocentre, the centre of mass of the Earth, and it is designed for positioning anywhere on Earth. The WGS84 datum is nothing more than a set of conventions, adopted constants and formulae. The WGS84 definition includes the following items:

1. The WGS84 Cartesian axes and ellipsoid are geocentric; that is, their origin is the centre of mass of the whole Earth including oceans and atmosphere.
2. The scale of the axes is that of the local Earth frame, in the sense of the relativistic theory of gravitation.
3. Their orientation, i.e. the directions of the axes and the orientation of the ellipsoid equator and prime meridian of zero longitude, coincided with the equator and prime meridian of the Bureau Internationale de l'Heure at the moment in time 1984 (that is, midnight on New Year's Eve 1983).
4. Since 1984 the orientation of the axes and ellipsoid has changed such that the average motion of the

crustal plates relative to the ellipsoid is zero. This ensures that the Z axis of the WGS84 datum coincides with the International Reference Pole, and that the prime meridian of the ellipsoid i.e. the plane containing the Z and X Cartesian axes, coincides with the International Reference Meridian.

5. The shape and size of the WGS84 biaxial ellipsoid is defined by the semi-major axis length meters and the reciprocal of flattening. This ellipsoid is the same shape and size as the GRS80 ellipsoid.

6. Conventional values are also adopted for the standard angular velocity of the Earth, and for the Earth gravitational constant. The first is needed for time measurement and the second to define the scale of the system in a relativistic sense.

There are a few more considerations, which are essential in understanding WGS 84, completely. First of all the ellipsoid is designed to best fit the Geoid of the Earth as a whole. This means that it generally doesn't fit the Geoid in a particular country as well as the non-geocentric ellipsoid used for mapping that country.

Secondly, the axes of the WGS84 Cartesian system - lines of latitude and longitude in the WGS84 datum, are not stationary with respect to any particular country. Due to tectonic plate motion, different parts of the world move relative to each other with velocities of the order of ten centimetres per year. The International Reference Meridian and Pole and the WGS84 datum, are stationary with respect to the average of all these motions. But this means they are in motion relative to any particular region or country. Over the course of a decade or so, this effect becomes noticeable in large-scale mapping. Some parts of the world e.g. Hawaii and Australia are moving at up to one metre per decade relative to WGS84.

4.7.3 Local Coordinate Systems

Just as with GPS coordinates, local coordinates or coordinates used in a particular country's maps are based on a local ellipsoid, designed to match the geoid in the area. Usually, these coordinates will have been projected onto a plane surface to provide grid coordinates. The ellipsoids used in most local coordinate systems throughout the world were first defined many years ago, before the advent of space techniques. These ellipsoids tend to fit the area of interest well but could not be applied to other areas of the Earth. Hence, each country defined a mapping system/ reference frame based on a local ellipsoid. When using GPS, the coordinates of the calculated positions are based on the WGS84 ellipsoid. Existing coordinates are usually in a local coordinate system and therefore the GPS coordinates have to be transformed into this local system.

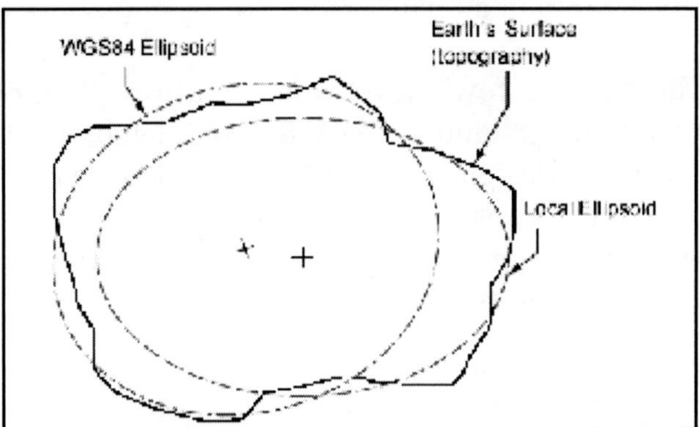

Fig. 4.7: Ellipsoids and Earth's Surface

4.7.3.1 WGS 84 and Everest

WGS-84 Coordinate System is being adopted universally as the standard form of Geographical Coordinates Representation System. This system has come into existence only towards the end of 20th Century, and prior to that the local coordinate system, Everest or Indian System, had been in use in India for more than 150 years.

A mathematical spheroid roughly representing the shape of Indian sub-continent has been assumed by the

Survey of India and all measurements are related to this spheroid. The reference datum fixed by Survey of India is located near Kalianpur in Madhya Pradesh. This is known as 'Everest-1830'. Everest is the name of first Surveyor General of India 'Sir George Everest' and the term '1830' represents the year in which the spheroid was defined. This datum is also sometimes called as 'Indian Datum'. The Indian spheroid has been marginally modified on a number of occasions so that the parameters assumed for the spheroid have been refined slightly from time to time. For example, such changes were made in the year 1930, 1956 and so on. Thus sometimes the Everest spheroid or Indian spheroid is also referred as Indian-1880, Everest-1930 or Indian-1956 etc. Even now some changes to the original definition are under consideration by Survey of India. Most of the maps, records and data available in India are in Everest system only. Till such time all the records are transformed to WGS-84 and new maps are printed, the local system is also likely to continue along with WGS-84.

4.7.4 Geodetic Datum

Geodetic datums define the size and shape of the Earth and the origin and orientation of the coordinate systems used to map the Earth. In other words a geographical datum is a mathematical model representing the shape of the Earth that is used as a reference or starting point for the determination or calculation of latitude, longitude and height. Datums have evolved from those describing a spherical Earth to ellipsoidal models derived from years of satellite measurements. A *Geodetic Datum* consists of two parts:

1) A defined geodetic reference ellipsoid.
2) A defined orientation, position and scale of the geodetic system in space.

Modern geodetic datums range from flat-Earth models used for plane surveying to complex models used for international applications, which completely describe the size, shape, orientation, gravity field, and angular velocity of the Earth.

A three-dimensional Cartesian coordinate system is associated with every geodetic datum. This coordinate system must be fixed in the physical Earth. This specification of the origin and orientation of the coordinate system can be expressed in several ways. With local horizontal datums, these quantities were fixed by specifying the geodetic coordinates of an initial point and at least one azimuth. With the use of satellite geodesy, the origin and orientation of the coordinate system are determined by specifying the three-dimensional coordinates of a number of points. A coordinate system can also be specified by describing the relationship between it and another coordinate system. This is the case with NAD 83 and WGS 84.

Referencing geodetic coordinates to the wrong datum can result in position errors of hundreds of meters. Different nations and agencies use different datums as the basis for coordinate systems used to identify positions in geographic information systems, precise positioning systems, and navigation systems. The diversity of datums in use today and the technological advancements that have made possible global positioning measurements with sub-meter accuracies requires careful datum selection and careful conversion between coordinates in different datums.

4.7.5 Geoids and Ellipsoids

The Earth's physical surface is a tangible one encompassing the mountains, valleys, rivers and surface of the sea. It is highly irregular and not suitable as a computational surface. A more smoothed representation of the Earth is the Geoid. There are a number of definitions for this surface; a descriptive one is as follows: 'that surface that would be assumed by the undisturbed surface of the sea, continued underneath the continents by means of small frictionless channels.

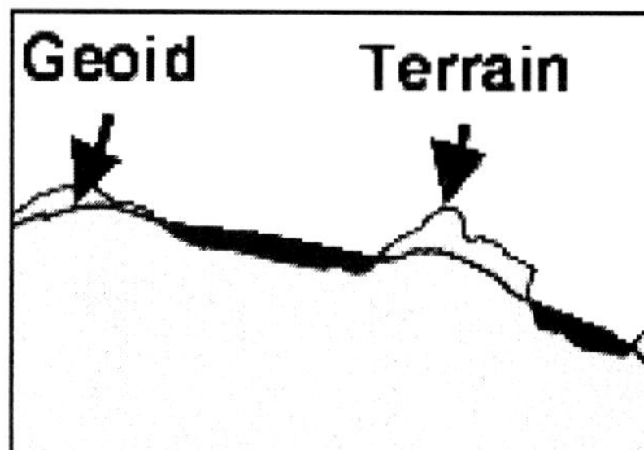

Fig. 4.8: Geoid and Terrain

The Ellipsoid is a smooth mathematical surface that best fits the shape of the geoid and is the next level of approximation of the actual shape of the Earth.

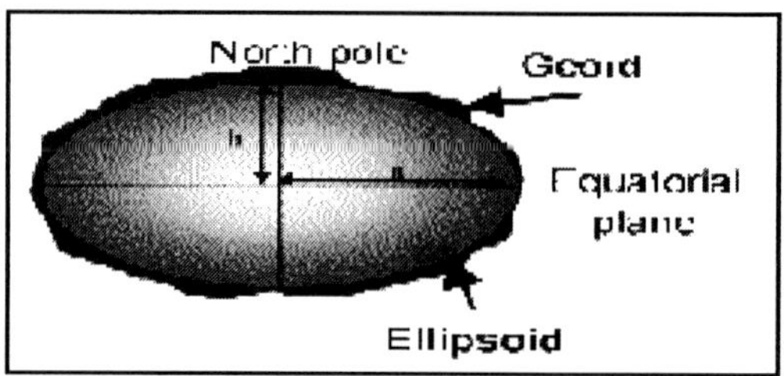

Fig. 4.9: Elements of an Ellipse

Elements of an Ellipse

a = Semi Major Axis

b = Semi Minor Axis

f = Flattening = (a-b)/a

PP′ = Axis of revolution of the Earth's ellipsoid

4.7.6 Ellipsoidal Co-Ordinates

The three dimensional i.e. real world, co-ordinates of a point on the Earth's surface can be defined in:

- Latitude (ϕ): angular displacement north/south of the equator.
- Longitude (λ): angular displacement east/west of the Greenwich meridian.
- Height : H) orthometric (height above mean sea level) or (h) ellipsoidal (Height above ellipsoid).

4.7.7 Projected Co-Ordinates

Plane co-ordinates are the simplest type of co-ordinates to use for everyday practical applications. To achieve this simplicity, the ellipsoidal latitude and longitude co-ordinates, or 3-D geocentric co-ordinates, are projected onto a plane surface. This is not possible without some distortion.

4.8 Errors, Accuracy and Precision

When carrying out any measurement it is important to quantify the "goodness" of the observation. For instance, if a position is to be determined with GPS, one should know a quantifiable degree of certainty, whether the observed position is accurate by 100 m or 10 cm. Accuracy refers to how close an estimate (or measurement) is to the true but unknown value, while precision refers to how close an estimate is to the mean estimate. It is possible to have high accuracy with low precision and vice versa. For example, in the figure bellow, the centre of the circles represents the "correct" position and each dot represents an individual estimate.

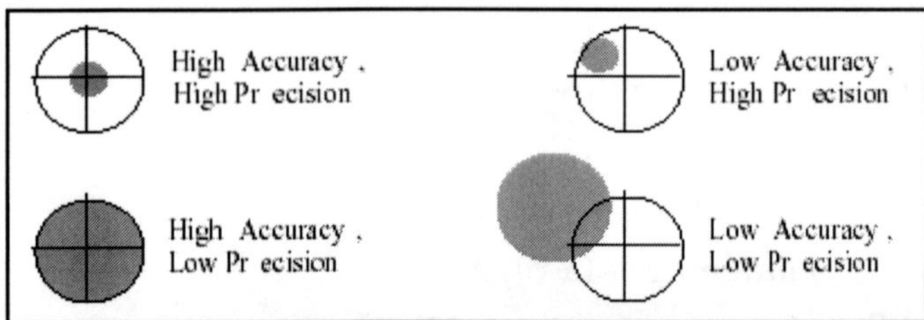

Fig. 4.10: Accuracy and Precision

4.8.1 Errors

The errors which limit the accuracy of any measurement may be classified as gross, systematic or random. Gross errors (other wise known as blunders) are errors which result from some equipment malfunction or observer's mistake. For example, if an operator of a GPS receiver recorded the height of an antenna above a monument as 0.5 m instead of a correct height of 1.5 m a gross error is said to have occurred. Gross errors must be detected and corrected. Systematic errors are those which have some known pattern or behavior which biases the observations. Ideally systematic errors are removed from observations by modeling. If all gross and systematic errors are removed from observations, only random errors remain. Precision includes only random effects, while accuracy includes both random and systematic effects. Random errors have the property that if enough observations are made there will be equal probability of negative and positive errors, yielding a mean value of zero. Random errors, according to statistical theory, tend to be distributed about the mean following the normal probability distribution function.

4.8.2 *Absolute and Relative Accuracy*

Similar to absolute and relative GPS positioning accuracy measures may also be considered as absolute or relative. Absolute accuracies are estimates of how close a position is to the truth in the Earth's reference frame. And relative accuracies are estimates of how well a vector between two points is measured (e.g. the accuracy of a distance measurement between two points).

Absolute accuracies are always represented as constant values. Relative accuracies may be represented as constant values, as parts per million (ppm), or both. Parts per million are used to relate error magnitudes with baseline length. For example 1 ppm corresponds to a 1 mm error over 1 km and a 1 cm error over 10 km.

4.8.3 *Error Sources*

There are several sources of error that degrade the GPS position from a few meters to tens of meters. These error sources are:

a. Ionospheric and atmospheric delays
b. Satellite and receiver clock errors
c. Multipath
d. Dilution of precision

e. Signal-to- noise ratio
f. Selective availability (S/A)
g. Anti spoofing (A-S)

4.8.4 Ionospheric and Atmospheric Delays

The ionosphere is a layer or layers of ionized air surrounding the Earth extending from almost 80 km above the Earth's surface to altitudes of 1000 km or more. The air is extremely thin at these altitudes. When the atmospheric particles are ionized by radiation (e.g., ultraviolet radiation and X-rays from the Sun), they tend to remain ionized due to few collisions between free negatively charged electrons and positively charged atoms and molecules called ions. These ions characterize the ionosphere. The free electrons affect the propagation of radio waves, thus the GPS signals. Unlike the troposphere, the ionosphere is a dispersive medium for radio waves, which means that the modulations on the carrier and carrier phases are affected differently and this effect is a function of carrier frequency. The impact decreases with the increased frequency.

Normally, the radiated energy from a transmitter gets through the ionosphere, in part absorbed by the ionized air, and in part refracted or bent downward again towards the Earth's surface. Further, carrier frequencies below about 30 MHz are reflected by the ionosphere, thus only higher frequencies, such as GPS signals, television and frequency-modulation (FM) radio, can normally penetrate the ionosphere. The ionosphere can be divided into two main layers called the E layer (about 80 to 113 km.) and the F layer (which is above the E layer). The E layer reflects low frequency radio waves while the F layer reflects higher-frequency radio signals. The F layer is composed of two layers: the F1 and F2 layers, which start approximately at 180 and 300 km above the Earth's surface, respectively. The thickness of the F layer changes at night, thus altering its reflecting characteristics.

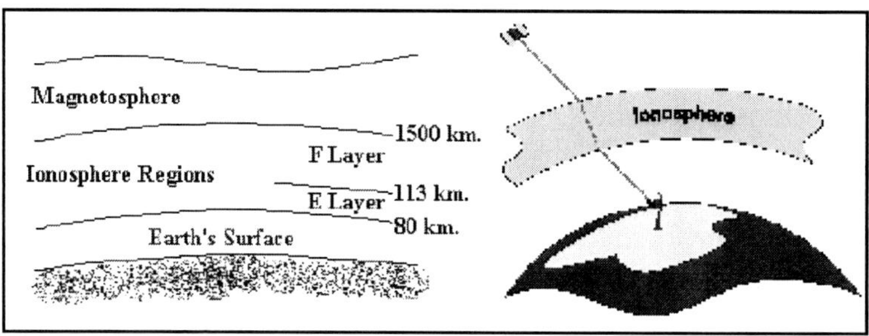

Fig. 4.11: Ionosphere

As the satellite signal passes through the ionosphere, it can be slowed down, the effect being similar to light refracted through a glass block. These atmospheric delays can introduce an error in the range calculation as the velocity of the signal is affected. (Light only has a constant velocity in a vacuum). The ionosphere does not introduce a constant delay on the signal. There are several factors that influence the amount of delay caused by the ionosphere, a few of them are as follows:

(i) Satellite elevation: - Signals from low elevation satellites will be affected more than signals from higher elevation satellites. This is due to the increased distance that the signal passes through the atmosphere.

(ii) The density of the ionosphere is affected by the sun. At night, there is very little ionospheric influence. In the day, the sun increases the effect of the ionosphere and slows down the signal.

(iii) Water Vapour also affects the GPS signal. Water vapor contained in the atmosphere can also affect the GPS signal. This effect, which can result in position degradation, can be reduced by using atmospheric models.

The user equalent range error (UERE) represented by the ionosphere delay is variable and can range from about ±5m to as great as ±150m at the receiver.

The effects of the troposphere (surface to about 50 km) on the signal are similar to those of the ionosphere in that the density of the troposphere causes s refraction of the signal from a straight-line path to the receiver. However, the effects of the troposphere are not as sever as those of the ionosphere. The user equalent range error (UERE) represented by the troposphere delay is variable and can range from about ± 2m to ± 20 m at the receiver.

4.8.5 Satellite and Receiver Clock Errors

The ability of a GPS receiver to determine a fix depends on its ability to determine how long it takes a signal to get from the satellite to the receiver antenna. This requires that the atomic clocks in the satellite be synchronized. Even a small amount of difference in the clocks can make a huge difference in the distance measurements. Ground stations throughout the world which monitor the satellites ensure that their atomic clocks are kept synchronized. These errors if not corrected by ground control stations can result in one meter errors. Receivers are also prone to timing errors. Receiver clock errors depend upon the oscillator provided within the unit. However, they can be calculated and then eliminated once the receiver is tracking at least four satellites.

4.8.6 Multipath

Multipath error result due to one or more reflected signals reach the receiver antenna in addition to the direct signal. Multi path occurs whenever the receiver antenna is positioned close to a large reflecting surface such as a lake

Fig. 4.12: Choke-Ring Antenna

or building. In such situation the satellite signal does not travel directly to the antenna but hits the nearby object first and is reflected into the antenna creating a false measurement.

Multipath can be reduced by the use of special GPS antennas that incorporate a ground plane (a circular, metallic disk about 50cm (2 feet in diameter) that prevent low elevation signals reaching the antenna. For highest accuracy, the preferred solution is use of a choke ring antenna. A choke ring antenna has 4 or 5 concentric rings around the antenna that trap any indirect signals. Multi path only affects high accuracy, survey type measurements. Simple handheld navigation receivers do not employ such techniques.

Multipath affects both, code and carrier observations. Multipath effects on position results can be minimized by observations over a longer time period. This is not possible in kinematics or rapid static surveying. Height error due to multipath is likely to be 15 cm. Multipath is difficult to detect and some times hard to avoid.

To minimize multipath effects the following measures may be adopted:

1. Select sites carefully and avoid nearby reflectors.
2. Use carefully designed antenna : Micro-strip, choke ring or cavity backed.
3. Use absorbing material near the antenna.
4. Setting elevation masks in the receiver to only accept satellites above a prearranged elevation (i.e., above 15°) can mitigate this.
5. Reject the signals that delay by too long a period of time.

4.8.7 Dilution of Precision

The Dilution of Precision (DOP) is a measure of the strength of satellite geometry and is related to the spacing and position of the satellites in the sky. The DOP can magnify the effect of satellite ranging errors. Different types of Dilution of Precision or DOP can be calculated depending on the dimension.

4.8.8 PDOP - Positional Dilution of Precision

The PDOP gives accuracy degradation in 3D position (vertical as well horizontal). GPS receivers usually report the quality of satellite geometry in terms of Position Dilution of Precision. PDOP refers to Horizontal (HDOP) and Vertical (VDOP) measurements (Latitude, longitude and altitude). A low DOP indicates a higher probability of accuracy, and a high DOP indicates a lower probability of accuracy.

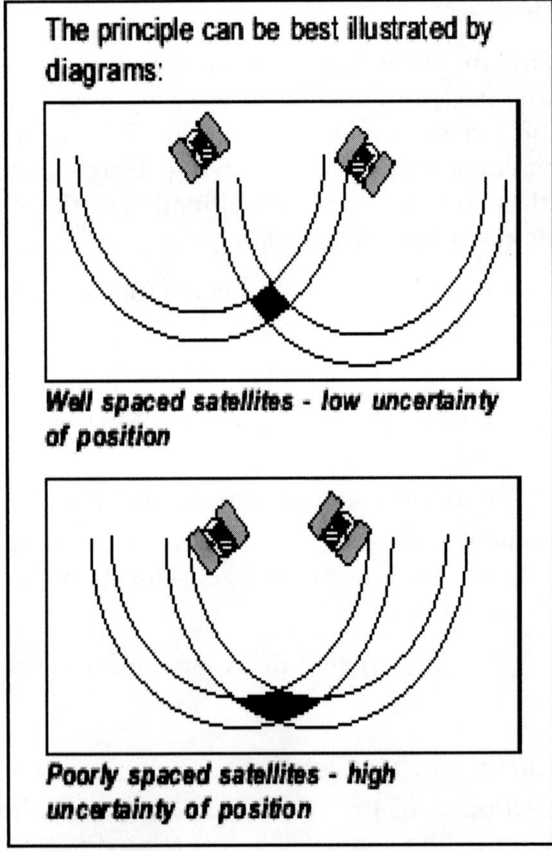

Fig. 4.13: Positional Dilution of Precision

4.8.9 GDOP - Geometric Dilution of Precision

Gives accuracy degradation in 3D position and time. The most useful DOP to know is GDOP since this is a combination of all the factors.

GDOP refers to where the satellites are in relation to one another, and is a measure of the quality of the satellite configuration. It can magnify or lessen other GPS errors. In general, the wider the angle between satellites, the better the measurement. Most GPS receivers select the satellite constellation that will give the least uncertainty, the best satellite geometry.

4.8.10 TDOP - Time Dilution of Precision

TDOP refers to satellite clock offset. The relationship between PDOP, TDOP, and GDOP is expressed through the following formulae:

$$GDOP = \sqrt{PDOP^2 + TDOP^2}$$

$$PDOP = \sqrt{HDOP^2 + VDOP^2}$$

4.8.11 Signal to Noise Ratio

This is a measure of the information content of a signal relative to the signal's noise. As this proportion decreases, information gets lost. With high SNR values (high signal, low background noise), fewer positions are required for averaging in order to get a confident position. Longer averaging times are required for lower SNR values. A value above 20 is good. The quality of a position is degraded if the signal strength of any satellite in the constellation is bellow 6.

4.8.12 Selective Availability (SA)

One of the largest sources of error in GPS has been selective availability (SA). Selective Availability is a process applied by the U.S. Department of Defense to the GPS signal. This is done by altering the satellite clock signals and modifying the orbital elements of the broadcast navigation code. This was implemented on March 25, 1990. This is intended to deny civilian and hostile foreign powers the full accuracy of GPS by subjecting the satellite clocks to a process known as 'dithering' which alters their time slightly. It is worth noting that S/A affects civilian users using a single GPS receiver to obtain an autonomous position. SA was turned off May 1, 2000. During its implementation, errors as high as ±100m were common. Users of differential systems are not significantly affected by S/A.

4.8.13 Anti spoofing (A-S)

Anti-Spoofing is similar to S/A ,it's intention is to deny civilian and hostile powers access to the P-code part of the GPS signal and hence force use of the C/A code which has S/A applied to it. Anti-Spoofing encrypts the P-code into a signal called the Y-code. Only users with military GPS receivers (the US and its allies) can de-crypt the Y-code.

Thus the accuracy of the position calculated from the GPS signal is subject to a number of biases and can be expressed by the equation:

$$P = \rho + c(dt - dT) + d_{ion} + d_{trop} + \check{Z}_p$$

Where:

P = calculated position

ρ = true position

c = constant

$(dt - dT)$ = clock error time difference

d_{ion} = ionospheric delay

d_{trop} = tropospheric delay

$\check{Z}_p$ = receiver noise and multipath

4.8.14 Blunders

Blunders can result in errors of hundred of kilometers. It may be due to Control segment mistakes due to computer or human error.

4.8.15 User mistakes

Includes incorrect geodetic datum selection, can cause errors from 1 to hundreds of meters. Receiver errors from software or hardware failures can cause blunder errors of any size.

The following table lists the possible sources of GPS error and their general impact on positioning accuracy:

GPS ERROR SOURCES

ERROR SOURCE	TYPICAL RANGE ERROR	DGPS (CODE) RANGE ERROR <100 KM REF-REMOTE
SV CLOCK	1 M	
SV EPHEMERIS	1 M	
SELECTIVE AVAILABILITY	10 M	
TROPOSPHERE	1 M	
IONOSPHERE	10 M	
PSEUDO-RANGE NOISE	1 M	1 M
RECEIVER NOISE	1 M	1 M
MULTIPATH	0.5 M	0.5 M
RMS ERROR	15 M	1.6 M
ERROR * PDOP=4	60 M	6 M

PDOP=Position Dilution of Precision (3-D) 4.0 is typical

4.9 Techniques to Improve GPS Accuracy

1. Number of visible satellites: When 3-dimensional coordinates are required (northern, easting, and elevation), a minimum of four satellites is needed. If only require 2-dimensional coordinates, at least three satellites are needed, but the 2 dimensional modes can significantly reduce the final accuracy.
2. Satellite elevations: When a satellite is low on the horizon, the satellite signals must travel a greater distance through the atmosphere, resulting in lower signal strength and delayed reception by the GPS receiver. Position data should be collected using only satellites that are at least 15° above the horizon.
3. Distance between base station and rover receivers: Accuracy degrades as the distance between base

station and rover increases. To obtain sub meter accuracy, capture data within 300 m of a base station. Using the satellite Differential GPS services (like Landstar, the subscription service for GIS Lab), this is not an issue—all of North America is within the baseline length limit for the network of base stations.

4. Occupation time at a point: Accuracy at a spot can be improved by collecting over a longer period of time at the location, then averaging these positions.
5. Differential GPS techniques: This involves using a second GPS receiver at a known location (a base station) to correct for the drift in GPS signals.

4.10 Differentially Corrected Positions (DGPS)

Many of the errors affecting the measurement of satellite range can be completely eliminated or at least significantly reduced using differential measurement techniques. DGPS allows the civilian user to increase position accuracy from 100m to 2-3m or less, making it more useful for many civilian applications. Differential GPS (DGPS) is a means of correcting for some system errors by using the errors observed at a known location to correct the readings of a roving receiver.

The basic concept is that the reference station "knows" its position, and determines the difference between that known position and the position as determined by a GPS receiver. This error measurement is then passed to the roving receiver which can adjust its indicated position to compensate.

The differential reference station computes the errors in the pseudo range measurements for each satellite in view separately, and broadcasts the error information, and other system status information, by some means. A differential beacon receiver receives and decodes this information, and sends it to the "differential ready" GPS receiver. The GPS receiver combines this information with the individual pseudo range measurements it makes, before calculating the position.

DGPS will eliminate the error introduced by Selective Availability, and errors caused by variations in the ionosphere, resulting in reported positions within about 10 metres of the true position 95 per cent of the time, for typical marine DGPS systems using inexpensive navigation receivers. Better receivers can get within 3 meters, or so. The DGPS correction data can be used as far as 1500 KM from the reference station depending on the DGPS setup – if the DGPS is part of a larger monitoring network.

A typical DGPS involves two types of receivers- the reference and rover receivers, using them the Differentially Corrected Positions are found out.

4.10.1 The Reference Receiver

The Reference receiver antenna is mounted on a previously measured point with known coordinates. The receiver that is set at this point is known as the Reference Receiver or Base Station. The receiver is switched on and begins to track satellites. It can calculate an autonomous position. Because it is on a known point, the reference receiver can estimate very precisely what the ranges to the various satellites should be. The reference receiver can therefore work out the difference between the computed and measured range values. These differences are known as corrections. The reference receiver is usually attached to a radio data link which is used to broadcast these corrections.

4.10.2 The Rover Receiver

The rover receiver is on the other end of these corrections. The rover receiver has a radio data link attached to it that enables it to receive the range corrections broadcast by the Reference Receiver. The Rover Receiver also calculates ranges to the satellites. It then applies the range corrections received from the Reference. This lets it calculate a much more accurate position than would be possible if the uncorrected range measurements were used. Using this technique, all of the error sources are minimized, hence the more accurate position. It is also worthwhile to note that multiple Rover Receivers can receive corrections from one single Reference.

The working principle of DGPS is easy to explain but in real life, it is a little complex. The following are the considerations which add complexity to this simple technique.

a. The Radio Link

There are many types of radio link that broadcast over different ranges and frequencies. The performance of the radio link depends on a range of factors including:

1. Frequency of the radio
2. Power of the radio
3. Type and 'gain' of radio antenna
4. Antenna position

b. The Type of Receivers

Networks of GPS receivers and powerful radio transmitters have been established, broadcasting on a "maritime only" safety frequency. These are know as Beacon Transmitters. The users of this service (mostly marine craft navigating in coastal waters) just need to purchase a Rover receiver that can receive the beacon signal. Such systems have been set up around the coasts of many countries. Other devices such as mobile telephones can also be used for transmission of data.

In addition to Beacon Systems, other systems also exist that provide coverage over large land areas operated by commercial, privately owned companies. There are also proposals for government owned systems such as the Federal Aviation Authority's satellite based Wide Area Augmentation System (WAAS) in the United states, the European Space Agency's (ESA) system and a proposed system from the Japanese government.

There is a commonly used standard for the format of broadcast GPS data. It is called RTCM format. This stands for Radio Technical Commission Maritime Services, an industry sponsored non profit organization. This format is commonly used all over the world.

4.11 Satellite Geometry

The satellite visibility and availability is an essential parameter for precise positioning and surveying. Users of GPS must know where, when and what satellites should be tracked to attain the best results. The following are the essential measures a GPS operator is expected to know, during positioning.

The elevation angle is the angle from the antenna between the horizontal and the line of sight to the satellite.

The Azimuth: is the clockwise angle from north to the location of the satellite in the sky.

Mask angle: the mask angle refers to the elevation angle below which GPS signals will not be recorded. A satellite is said to be visible if it is above the specified mask angle provided there is no obstructions are present. GPS receivers, processing software, or both may have an option to set a specific mask angle or cutoff angle.

Obstructions: Obstructions are objects which block the path between a satellite and receiver. For example, if a desired satellite is at an elevation of 20° and azimuth of 70°, and a building is located at the same elevation and azimuth, the satellite signal will be obstructed. The avoidance of obstructions is very important to the successful application of GPS positioning.

Almanac files: for any given location on the Earth and time, it is possible to predict which satellites will be available and their location in the sky. This is accomplished by using almanac files which contain satellite orbit parameters, in conjunction with software designed to use almanac files to compute satellite visibility.

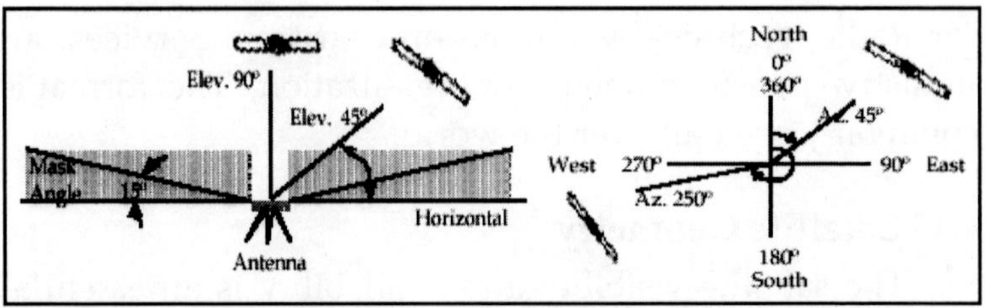

Courtesy: - Geomatics Canada, Information Services, 615 Booth Street, Ottawa, OntarioK1A 0E9

Software packages to compute satellite visibility are available commercially and often accompany commercial GPS receiver software. When computing satellite availability, one should be careful to use only recent almanacs, no more than one month old.

Satellite Geometry: The accuracy with which positions can be determined is not just a function of the measurement precision, and the appropriate modeling of biases. It is also a function of the satellite(s) - receiver(s) geometry.

Sky plots are generally used to represent satellite visibility. Each concentric ring represents an elevation angle, while each radiating line represents an azimuth. In the figure, the shaded area, below 15° elevation represents the mask angle.

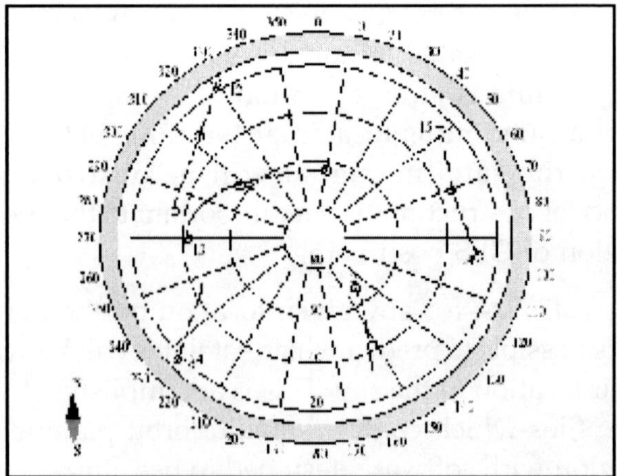

Fig. 4.15: A Sample Sky Plot

The path of all visible satellites over a two hour period is plotted. The numbers indicated on each plotted line are the satellite numbers. For example, satellite 13 is shown in the plot as tracing a path from an elevation angle of 40° and azimuth of 270° to an elevation angle of about 60° and azimuth of 10°.

Satellite geometry has a direct effect on positioning accuracies. The best single point positioning accuracies are achieved when satellites have good spatial distribution in the sky (e.g. one satellite overhead and the others equally spread horizontally and at about 20 ° elevations). Sub-optimal geometry exists when satellites are clumped together in one quadrant of the sky.

The effect of satellite configuration geometry is usually expressed by the Dilution of Precision (DOP) factor. DOP is the ratio of the positioning accuracy to the measurement accuracy:

$\sigma = DOP * \sigma o$

σo is the measurement accuracy, and

σ Is the position accuracy.

DOP is always a number greater than unity when there are no redundant observations. There are a number of different definitions of DOP factors, depending on the coordinate component, or combination of coordinate components, being considered:

$$PDOP = \sqrt{\sigma_E^2+\sigma_N^2+\sigma_H^2} = \sigma_X^2+\sigma_Y^2+\sigma_Z^2$$

$$HTDOP = \sqrt{\sigma_E^2+\sigma_N^2+\sigma_T^2}$$

$$HDOP = \sqrt{\sigma_E^2+\sigma_N^2}$$

$$VDOP = \sqrt{\sigma_H^2}$$

$$TDOP = \sqrt{\sigma_T^2}$$

Where:

σE^2, σN^2, σH^2	are the variances of the east, north and height components,
σx^2, σy^2, σz^2	are the variances of the X, Y and Z components, and
σT^2	is the variance of the estimated receiver clock error parameter

The geometry of satellites, as it contributes to positioning accuracy, is quantified by the geometrical dilution of precision (GDOP). Satellite configurations exemplifying poor and good GDOP are illustrated in the Figure 4.16.

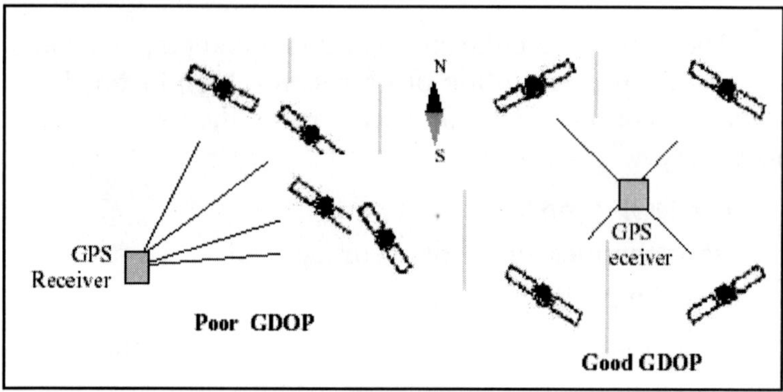

Fig. 4.16: GDOP

4.12 GPS Receivers and Software

There are quiet a different types of GPS receivers and Software available in the market. The choice of the GPS is highly depends on the requirements of user. The user must be clear about the following concepts before purchasing any GPS instrument or software from the market.

4.12.1 GPS Receivers

A receiver for the Global Positioning System must lock onto the signals from four of the GPS satellites to give a full three-dimensional position. More limited information can be gained from two or three satellites. Different types of receivers make use of different parts of the GPS Signal Structure. The following are the observable quantities by a GPS receiver,

The C/A code on carrier L1 (sufficient for +/-100 m positioning).

The P-code on carrier L1

The P-code on carrier L2

The phase of the L1 carrier

The phase of the L2 carrier

Uses of GPS receivers range from handheld positioning and personal navigation devices to precision surveying tools. Applications of the Global Positioning System require different types of receivers and make use of different parts of the GPS Signal Structure.

4.12.2 Handheld GPS Positioners

The most popular type of GPS receiver is a handheld positioning and navigation unit which monitors the C/A code on the L1 carrier of the GPS signal structure. Such units must be used outdoors with a clear view of the sky, and are capable of locking onto the signals from four or more GPS satellites. The signals allow them to calculate the distances to four satellites, and with that data they can calculate position on the Earth's surface in latitude and longitude within +/- 100 meters 95% of the time. They can measure altitude to an accuracy of +/- 180 meters. Since the four signals received are stabilized by atomic clocks, the timing accuracy of an ordinary digital watch is sufficient for the calculations done in the handheld unit. Waypoints may be set and stored in the units, and the received signal used to calculate range and direction to those waypoints. Speed and direction of travel can also be displayed.

4.12.3 Differential GPS

Position accuracies to about +/- 2 meters can be obtained by comparison of a field GPS receiver with a fixed receiver at a known position. Receivers for this kind of positioning differ from the popular hand-held units in that they have memories for storage of data in both the moving and stationary units and are calibrated to take data at the same time. The field data is then compared to the base station data in post processing to acquire the increased accuracy.

This accuracy is obtained using just the C/A code on carrier L1 of the GPS signal structure. To obtain position uncertainties under a meter in DGPS, receivers must measure both the C/A code and the phase of the L1 carrier. A sequence of phase measurements must be made to achieve the increase in accuracy. For such measurements to be made in real time, a radio link between the base and field stations must be used. A more accurate version of DGPS is called relative positioning GPS.

4.12.4 *Relative Positioning GPS*

A more sophisticated version of differential GPS is used by surveyors to obtain accuracies on the order of a centimeter for the difference in two positions. The GPS receivers for this process must record the C/A code and the phase of carrier L1 of the GPS signal structure at every epoch of the Navigation Message, on the order of every 12 to 15 seconds. The data from two receivers is post processed to obtain the sub-centimeter accuracy.

Maintaining this accuracy for distances over 30 km requires dual frequency receivers that measure the C/A code plus the phases of both carriers, L1 and L2. For both long lines and rapid measurement of shorter distances, receivers make use of all five of the measurable signals of the signal structure.

4.12.5 *Kinematic GPS*

Relative measurements "on the fly" may be made with kinematic packages on GPS receivers, so that accuracies comparable to the relative positioning GPS can be made while moving. Real time kinematic GPS requires a radio link between the field receiver and the base station, which are elaborately discussed in the chapter surveying with GPS.

4.12.6 *Software Packages*

Varieties of software packages are available for GPS phase data processing. These packages have either been developed by universities and government departments (for research uses, for internal operational purposes, or for very high precision scientific applications), or by the receiver manufacturers. Although the detailed architecture of GPS software varies considerably from one package to another, there are a number of functions that most comprehensive GPS surveying software packages must carry out.

The major components of any package include software to aid pre-survey planning, decision making and reconnaissance; to support field observations (including, increasingly, the high productivity survey methods such as "stop & go", "kinematic" and other such techniques), data pre-processing and checking; single-baseline data processing network adjustments and quality control, and the transformation of the results into an already established geodetic reference system. Various criteria can be used to evaluate software. The following is a list of some that may be appropriate.

Primary Criteria	**Secondary Criteria**
Accuracy	Level of automation
Reliability	Processing speed
Support high productivity surveying techniques	Processing flexibility
Efficiency of data flow	Data analysis and quality control capabilities
Ease of operation	Software portability and ease of maintenance
	Data file storage and I/O requirements
	Computer hardware requirements

Accuracy and reliability are dependent not only on the data processing algorithms that are used, but also on the planning and field procedures, as well as the hardware. Hence, if the field procedures result in the collection of poor quality data, or an inadequate quantity of data, then no matter how sophisticated the software, the accuracy and reliability of the baseline results will suffer. However, accuracy (and to a lesser extent reliability) do directly influence the selection of the mathematical models and the processing strategies to be employed for any given survey application.

Although efficient and fast processing is desirable, this will obviously be dependent upon the computer implementation, the algorithms used, the amount of data to be processed, and the accuracy required. High accuracy applications generally require the flexibility to intervene at various stages to control the processing. For scientific or geodetic applications, commercial software is far from adequate. "Part per million" accuracy (standard surveying) applications place a higher priority on automation or ease-of-use, with minimal operator intervention.

Nowadays, an important marketing factor for a GPS "package" is its ability to support high productivity survey techniques. The field procedures are very different to those of the conventional static GPS surveying method. However, it is the innovative data processing algorithms that make possible the generation of high accuracy results in a fraction of the time required by the conventional GPS baseline surveying technique.

Efficient data flow implies a natural flow of data from raw to finished form in some logical and efficient sequence of operations, and consequently has a direct impact on the software structure. Once software architecture has been specified, it then becomes possible to address the related secondary criteria of automation, data storage, and computer hardware requirements, etc.

It is also important that the software structure allow for flexibility in processing options, and be adaptable to changes as new models and methodologies are developed. This is particularly the case for real-time positioning implementations, or where centimetre-level positioning is required for mission critical applications such as machine guidance and control.

Finally, the software should be portable to allow easy implementation in different computing environments. This may be difficult to achieve if the user interface is device dependent, and requires significant operator input. Hence the software structure, the operating system requirements, and data management structure selected will require a similar trade-off having to be made between efficiency, speed, and automation on the one hand, and software portability and maintainability on the other.

Chapter - 5

SURVEYING WITH GPS

5.1 Introduction

The determination of the exact boundaries and location of a property is what is known generally as surveying. Some of the first known surveyors were Egyptian surveyors who used distant control points to replace property corners destroyed by the flooding in Nile River. Later, the Greeks and Roman surveyed their settlements. The Dutch surveyor Snell Van Royen was the first who measured the interior angles of a series of interconnecting triangles of the baselines to determine the coordinates of points long distances apart. The surveyor's role in society has remained unchanged from the earliest days; that is to determine land boundaries, provide maps of his environment, and control the construction of public works. The chain of technical developments from the early astronomical surveyors to the present satellite geodesists reflects man's desire to master time and space and use science to further his society.

Surveying with GPS has become popular due to the advantages of accuracy, speed, versatility and economy. The techniques employed are completely different from the classical surveying methods. Certain basic rules are followed in GPS surveying, which relatively produce good results.

5.2 GPS and Survey

The high precision of GPS carrier phase measurement, together with appropriate adjustment algorithms, provides an adequate tool for a variety of tasks for surveying and mapping. Using DGPS methods, accurate and timely mapping of almost anything can be carried out. The GPS is used to map cut blocks, road alignments and environmental hazards such as landslides, forest fires, and oil spills and for cadastral mapping, continuous kinematic techniques can be used for topographic surveys and accurate liner mapping.

Merits and Demerits of using GPS for Survey

The following are the merits and demerits of using GPS for survey:

Merits

1. Intervisiblity not required.
2. Operations are not weather independent.
3. Geodetic accuracies easily achieved.
4. 3-D coordinates are obtained

Demerits

1. High capital cost of GPS instrumentation.
2. GPS surveying is generally "targeted" to satisfy a specific survey need.
3. GPS cannot be used in indoor.

5.3 Elements of the GPS Survey

The elements of the GPS survey includes, Definition of the task, planning and preparation; field operations; data processing; and final reporting. Validation and reconnaissance form an integral part of the planning and preparation phase.

5.3.1 Definition of the Task

Before going to the field one should clearly define the following:

1. Number of points need to be collected,
2. Accuracy requirement in horizontal and vertical positions and the nature of points distribution,

3. Knowledge about the local area, like: how far it is, what is the near by town if the area to be surveyed is a village, language spoken, food, water availability in that area, etc.

5.3.2 *Planning and Preparation*

Planning and preparation for a GPS field project begins with the identification of positioning requirements and ends with complete readiness for successful field operations. The extent of all the intermediate steps varies greatly with the magnitude, accuracy and locality of the project. As a preliminary step, the points to be positioned and their accuracy requirements should be identified. Then, the sites to be positioned and the available survey control should be plotted on a map, Topographical maps at 1:50,000 and 1:250,000 are well suited for this purpose. The following are the important steps in the planning and preparation phase:

a) Selection of the right positioning technique,

b) Selection of receiver type,

c) Validation, reconnaissance,

d) Survey design and preparations

5.3.2.1 Selecting the Right Technique

There are many aspects which influence the choice of positioning technique. Accuracy requirements, the geographical environment, the distance between points to be positioned and the costs are major considerations.

5.3.2.2 Selecting the Receiver Type

GPS receivers may be either leased or bought. Whatever the case, it is suggested that all receivers used together for relative positioning be of the same make to avoid problems which often result from mixing receiver types such as biases, complexities in data processing and data rate incompatibilities.

For semi-kinematic, rapid static and conventional static GPS surveys, code and carrier measurements are required. For short baselines using conventional techniques, single frequency receivers are sufficient. For conventional static GPS over longer baselines where high accuracies are sought, dual frequency receivers are desirable since they permit correction of most of the ionospheric errors. For rapid static surveys, dual frequency receivers are strongly recommended since they enable sophisticated data handling methods for ambiguity resolution. The following table explain different GPS methodologies suitable for code and carrier measurements.

Methodology	GPS Measurements Required
Single Point Positioning	code
Code Differential	code
Semi-kinematic	code & carrier
Rapid Static	code & carrier, dual frequency preferable
Conventional Static	code & carrier, dual frequency for long baselines

GPS Measurements Required for Varied Positioning Techniques.

5.3.2.3 Validation

In the planning phase of a GPS project the procedures and equipment to be used, from data collection to the final product, should be tested to ensure that they reliably satisfy the desired accuracy requirements. This testing is referred to as the validation process. The following are the three main components tested in the validation process:

- The positioning technique chosen,
- The equipment to be used and,
- The processing method adopted.

By testing them it is easy to identify and solve problems related to either of the technique or of the equipment. Further, validation is highly useful in determining the expected accuracies.

Validation is an important evaluation and feedback step in the project plan. An equally important step, which gives feedback in the planning process, is field reconnaissance.

5.3.3 *Reconnaissance*

Reconnaissance consists of checking field project sites before commencing GPS observations. Sites should be checked for their suitability for GPS, availability of control, and logistical requirements. The final product of field reconnaissance will include a set of points ready for GPS observations as well as a description for each site, access information and a description of any special steps, which need to be taken.

5.4 Survey Design

Another important step in the planning and preparation process is the survey design. Considerations in the survey design include control requirements, network configuration and redundancy. Obviously, the survey design will vary greatly depending on the accuracy sought and the GPS positioning technique employed.

5.4.1 *Preparations*

The following are the necessary preparations before one venture into the field for collecting positions:

1. Decide the optimal number of GPS receivers and personnel for the project and make the necessary arrangements.
2. Plan the survey design, taking into account control requirements, network configuration, and travel time between sites, satellite window and logistical constraints, such as traveling route, etc.,
3. Establish a unique numbering or naming system to clearly identify all sites positioned on the ground with

their related computer data files, positional information and other associated attributes.
4. Arrange for transportation between sites (e.g. car, helicopter, boat, or foot).
5. Train personnel on receiver operation, GPS observing procedures and data processing.
6. Organize accommodations in the field if required.
7. Organize all required equipment and supplies to support GPS field activities.

5.4.2 Field Operations

Responsibilities in the field are typically divided amongst a party head, observers and a processor. Depending on the magnitude and methodology of the project, these three groups of responsibility may be assigned to one person or shared amongst many.

The party chief should monitor the schedule of observations as per plan, check for satellite problems, geomagnetic storms, assess daily results and modify plans whenever required and handle any logistical difficulties.

The observers in-turn check that they have all the required equipment, should ensure receiver battery is fully charged before going to the field, allow enough time to travel to the site, verify correct station level, centre and orient GPS antenna, measure antenna height- incase height measurement is required, initialize receiver, monitor receiver operational and data recording, complete data log sheet and submit data and log sheets to the processor after collecting positions and measurements.

The processor generally verify the data submitted to them, down load the data from palm top or other hand held devices, make backups of the raw data, organize all the data, process GPS data, check results and report to the party head.

5.4.3 Data Processing and Final Reporting

The data processing and final reporting are not as critical as planning, preparation or field operations, but still must be given due attention for the overall success of the project.

The complexity of data processing corresponds with the complexity of the GPS technique used. Single point positioning is the simplest, followed by differential positioning and then carrier techniques. Most single point positioning solutions are computed within the receiver and displayed. The only post-processing activities, involve downloading the data and combining it in a database or a geographically referenced information system.

For differential solutions using code observations, the data from the monitoring site and all the rover sites must be loaded onto one computer. At the start of the processing program, the known monitor receiver NAD83 coordinates should be entered. The program will then match the times of the code observations made at each remote site with those made at the monitor site. By using the satellite ephemeris data, the known receiver coordinates and the code measurements, the program will compute the coordinates for each remote site.

Processing for conventional static GPS surveys is more complex and may require combining several sessions of observations. All data for one session must be loaded onto a computer. As well, the appropriate "known" three-dimensional NAD83 coordinates of the control points should be entered in the processing program. Most software will also require that approximate coordinates for all other sites occupied during the session be entered. Such approximate values may be read off the same receiver used in the field, or may be scaled from a map. For each session processed, most software will require one point be held fixed three dimensionally. Ideally this point will be a control point with known NAD83 coordinates. If a control point is not included in a specific session, coordinates of a site in common with an adjoining session that was tied to a control point should be used.

In the GPS processing algorithm, models may be used to correct some of the biases in the observations. Then all the observations, the ephemeris data and the known coordinates will be combined together in an optimal way (know as an adjustment) to arrive at a solution. Techniques for processing semi-kinematic and rapid static data are still evolving and have similarities with both differential processing and conventional static processing.

When all field operations and observations are complete, a final report of the project should be drafted, documenting the stations occupied, methods used and results attained. Perhaps the most important requirement for final reporting is the identification of any improvements which could be made in the procedures, for the next GPS positioning project.

5.5 GPS Measuring Techniques

To the surveyor or engineer more important than the theory behind GPS, are the practical application and the effective use of GPS. Like any other tool, GPS is only as good as its operator. Proper planning and preparation are essential ingredients of a successful survey, as well as an awareness of the capabilities and limitations of GPS. There are several measuring techniques that can be used by most GPS Survey Receivers. The surveyor should choose the appropriate technique for the application. There are two broad categories of GPS measurements. Positioning with GPS may take the form of either single point positioning or relative positioning.

5.5.1 Single Point Positioning

In single point positioning coordinates of a receiver at an "unknown" point are sought with respect to the Earth's reference frame by using the "known" positions of the GPS satellites being tracked.

Single point positioning is also referred to as absolute positioning, and often just as point positioning. In relative positioning the coordinates of a receiver at an "unknown" point are sought with respect to a receiver at a "known" point. The concept of single point positioning is illustrated in Figure bellow. Using the broadcast ephemerides, the position of any satellite at any point in time may be computed.

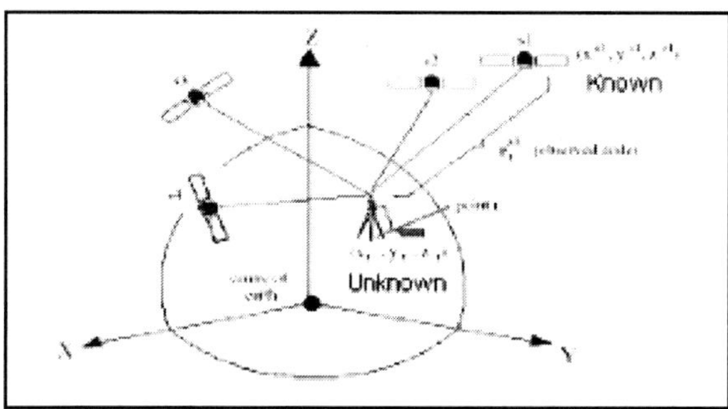

Fig. 5.1: Single Point Positioning

5.5.2 Relative Positioning

The concept of relative positioning is illustrated in the Figure bellow. Instead of determining the position of one point on the Earth with respect to the satellites (as done in single point positioning), the position of one point on the Earth is determined with respect to another "known" point. The advantage of using relative rather than single point positioning is that much higher accuracies are achieved because most GPS observation errors are common to the known and unknown site and are reduced in data processing. There are a number of relative positioning methods, which include Static survey, Rapid static survey, Kinematic survey and Real time kinematic (RTK) survey.

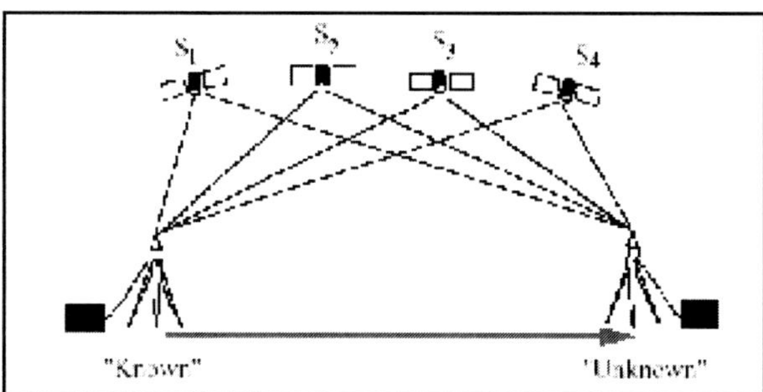

Fig. 5.2: Relative Positioning

5.5.2.1 *Static Survey*

This was the first method to be developed for GPS surveying. It can be used for measuring long baselines (usually 20km and over), geodetic networks, tectonic plate studies etc. This offers high accuracy over long distances but is comparatively slow and time consuming.

The principle is that 'One receiver is placed on a point whose coordinates are known accurately in WGS84. This is known as the Reference Receiver. The other receiver is placed on the other end of the baseline and is known as the Rover. Data is then recorded at both stations simultaneously. It is important that data is being recorded at the same rate at each station. The data collection rate may be typically set to 15, 30 or 60 seconds.

The receivers have to collect data for a certain length of time. This time is influenced by the length of the line, the number of satellites observed and the satellite geometry (dilution of precision or DOP). Once enough data has been collected, the receivers can be switched off. The Rover can then be moved to the next baseline and measurement can once again commence. It is very important to introduce redundancy into the network that is being measured. This involves measuring points at least twice and creates safety checks against problems that would otherwise go undetected. A great increase in productivity can be realized with the addition of an extra Rover receiver. Good coordination is required between the survey crews in order to maximize the potential of having three receivers.

5.5.2.2 *Rapid Static Surveys*

Rapid Static surveys, used for establishing local control networks, Network densification etc, offer high accuracy on baselines up to about 20km and are much faster than the Static technique. Here, a Reference point is chosen and one or more Rovers operate with respect to it. Typically, Rapid Static is used for densifying existing networks, establishing control etc. When starting work in an area where no GPS surveying has previously taken place, the first task is to observe a number of points, whose coordinates are accurately known in the local system. This will enable a transformation to be calculated and hence, all points measured with GPS in that area can be easily converted into the local system, at least four known points on the perimeter of the area of interest should be observed. The transformation calculated will then be valid for the area enclosed by those points. The Reference receiver is usually set up at a known point and can be included in the calculations of the transformation parameters. If no known point is available, it can be set up anywhere within the network. The Rover receiver(s) then visit each of the known points. The length of time that the Rovers must observe for at each point is related to the baseline length from the Reference and the GDOP. The data is recorded and post-processed back at the office.

5.5.2.3 Kinematic

The Kinematic technique is typically used for detail surveying, recording trajectories etc., although with the advent of RTK its popularity is diminishing. The technique involves a moving Rover whose position can be calculated relative to the Reference.

At first, the Rover has to perform what is known as an initialization. This is essentially the same as measuring a Rapid static point and enables the post-processing software to resolve the ambiguity when back in the office. The Reference and Rover are switched on and remain absolutely stationary for 5-20 minutes, collecting data. (The actual time depends on the baseline length from the Reference and the number of satellites observed). After this period, the Rover may then move freely. The user can record positions at a predefined recording rate, can record distinct positions, or record a combination of the two. This part of the measurement is commonly called the kinematic chain. A major point to watch during kinematic surveys is to avoid moving too close to objects that could block the satellite signal from the Rover receiver. If at any time, less than four satellites are tracked by the Rover receiver, one must stop, move into a position where 4 or more satellites are tracked and perform an initialization again before continuing.

5.5.2.4. Real Time Kinematic (RTK) Survey

RTK stands for Real Time Kinematic. It is a Kinematic on the Fly survey carried out in real time. RTK surveying requires two receivers, recording observations simultaneously, and allows the rover receiver to be moving. RTK surveying techniques also use dual-frequency LI/L2 GPS observations and can handle loss of satellite lock. The RTK technology allows the rover receiver to initialize and resolve the integer ambiguities without a period of static initialization. With RTK, if loss of satellite lock occurs, initialization can occur while in motion. The integers can be resolved at the rover within 10-30 sec, depending on the distance from the base station.

RTK surveying requires dual frequency LI/L2 GPS receivers. One of the GPS receivers is set over a known point, while the other receiver may be free to travel from point to point. If the survey is performed in real time, a radio link and a processor or data collector are needed. The radio link is used to transfer the raw data from the reference station to the rover. RTK surveys can be accurate to within 0.05 to 0.10 feet (2 – 3 centimeters), providing a good static network and calibration were performed prior to performing the RTK survey.

Most RTK GPS systems make use of small UHF radio modems. Radio communication is the part of the RTK system. It is worth considering the following influencing factors when trying to optimize radio performance:

1. Power of the transmitting radio, if more, the better will be the performance. However, most countries legally restrict output power to 0.5 - 2W.
2. Height of transmitter antenna is another important factor. Radio communication can be affected by line of

sight. The higher up the position of the antenna, the less likely to get line of sight problems. It will also increase the overall range of radio communication. The same also applies to the receiving antenna.

Other influencing factors affecting performance include the length of the cable to radio antenna (longer cables mean higher losses) and the type of radio antenna used.

5.5.2.5 Comparison between Static and Kinematic Survey

In static positioning, a GPS receiver is required to be stationary whereas in kinematic positioning a receiver collects GPS data while moving. The concepts of static and kinematic positioning for both single point and relative positioning cases are illustrated in the Figure bellow. Note that for kinematic relative positioning one receiver, referred to as a monitor, is left stationary on a known point while a

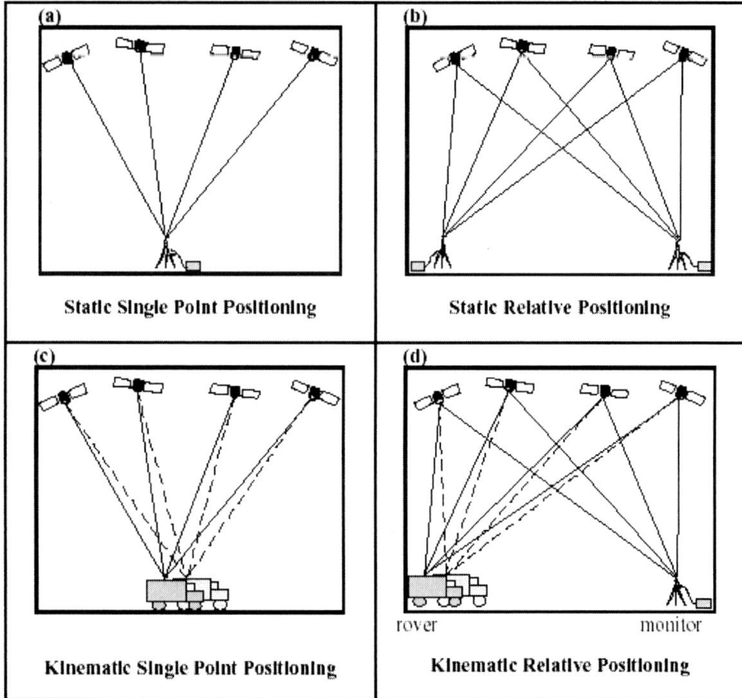

Fig. 5.3: Static and Kinematic Survey

second receiver, referred to as a rover, is moved over the path to be positioned.

5.6 Tips for Successful GPS Survey

For successful, high-accuracy GPS surveying it is advisable to take the observations in good windows.

Window for Rapid Static Survey	**Window for Other Survey**
4 or more satellites above	5 or more satellites.
GDOP 8.	GDOP 5.
Satellites above 15° cut-off angle.	Satellites above 20° cut-off angle

Always

1. Use sky plot to check for obstructions.
2. Re compute GDOP if a satellite is obstructed.

For Static and Rapid Static surveys, always fill out a record sheet for each point you survey. An example is given on the next page. With Static and Rapid Static surveys, it is vital that the antenna height is measured correctly. This is one of the most common mistakes when carrying out a GPS survey. Measure the height at the beginning and end of occupation. With Kinematic and RTK Surveys, the antenna is usually mounted on a pole which has a constant height. During Static and Rapid Static surveys, the GPS antenna has to be kept totally still. This also applies to the Rapid Static initialization of Kinematic surveys but not to Kinematic on the Fly or RTK surveys.

Chapter - 6

INTEGRATION OF GPS AND GIS

6.1 Geographic Information System (GIS)

GIS, an acronym of Geographical Information System, is an information system used for management decisions and allows one to create, maintain and query electronic databases of information normally displayed on maps. These databases are spatially-oriented, the fundamental integrating element being the position on the Earth's surface. GIS consists of a set of computerized tools and procedures that can be used to effectively encode, store, retrieve, overlay, correlate, manipulate, analyze, query, display (both graphically and numerically) and disseminate earth related information. The main difference or factor which separates GIS from other information storage and retrieval systems is the spatial and specific locational features given in a co-ordinate form, for reference and use as an important variable in quantitative analysis.

The spatial data is generally in the form of maps, which could be showing topography, geology, soil types, forest and vegetation, land use / land cover, water resources, GPS reading, etc., and stored as layers in a digital form in the computer. Besides spatial data, one could also have attribute data, like statistics, texts and tables. New maps can be easily generated by integrating many layers of data in a computer. Thus, a GIS has a database of multiple information layers that can be manipulated or analyzed to evaluate relationships among the selected elements, from all that is available.

The power of GIS lies in its ability to identify relationships between features based on their locations and their attributes. GIS allows one to view these relationships in maps, charts, tabular forms etc. In GIS one has the ability to create different layers of data. For example, one can create layers for roads, traffic signs, buildings, fire hydrants, water lines, etc and once the layers are created one can superimpose different layers and analyze the relationship between various features and their attributes. In order to analyze the data one requires relevant and current information, which can be obtained from various sources. Some data can be obtained from existing sources e.g., census, etc. However, other data specific to a project may not be readily available and may have to be collected for a specific project with GPS. Once the data is collected in GPS receivers, it can be directly transferred to a GIS System. This data will have the spatial location as well as feature/ attribute data attached to it.

6.2 Integrating – GPS/GIS Technology

Combining the GPS data with GIS allows far greater capabilities than what GPS and GIS can provide individually. With the combination of two technologies one is able to display the "FIELD/ACTUAL SITE" on a computer and make informed decisions. There is no need to make specific site visits or review several documents/ drawings. Also, another benefit of the integration is the fact that the data can be shared by unlimited users in various departments for their own specific needs and analysis.

Another important advance in this technology has been the introduction of software which allows bringing into GIS not only GPS position information but a digital picture (satellite image) of the same area. With the software, one can study relationships between features and also view actual photographs of the features on the computer.

6.3 Pitfalls of Integration

Combination of GPS/GIS technology is limited by the amount of data. As explained earlier one needs lot of good data for conducting analysis. While some data is available, a lot of data has to be generated by the users for their use. Sometimes collection of field data could turn out to be time consuming and expensive.

Data collected must be accurate and meet the correct formats. For example one need to make sure that all the "layers" of data displayed are in same units (feet/meters) and the projections and datums match. Without this the analysis will not be correct. However, the integration of GPS with GIS brings the real world to the desktop. What could take days to visit a specific site and analyze can now be performed on the desktop.

6.4 Process of Integration

Once GPS readings were available they need to be processed further before being converted into desired format such as maps. Both GPS and GIS are systems executed through software, all the processes of integration have to flow through them. There is a variety of GIS software available in the market, viz., Arc Map, Arc View, Geomedia, etc. Some GPS instruments like the products from Leica are executed through the software Arc Pad - A Product of ESRI. The processes of integration thus involve the modules of both GPS and GIS software. The following are the major processes involved in the integration.

1. Editing & Validation
2. Transformation
3. Projection
4. Mapping and Map Layout

6.4.1 Editing and Validation

There are generally two different types of errors could be anticipated from a GPS reading, viz., errors related to the position of the point, errors related to linear feature.

Point positioning errors may be due to varied reasons, discussed elsewhere in this book. These errors related to points could be solved by precise positioning techniques, observing the same point at different time periods and satellite availability.

Errors related to linear features are of two types, viz., over shoot and Undershoot. Overshoot is the line gone beyond its target snap position. While Undershoot is related to the line which is felt short of some distance before its target snaps position. Overshoot could be corrected by trimming the extended line while the undershoot could be snapped to the target snap position by extending the line segment.

6.4.2 Transformations

The purpose of a transformation is to transform coordinates from one system to another. There are several different transformation approaches exist. The choice of the transformation approach lies in the accuracy requirement. However, the procedure for the determination of transformation parameters is same for all approaches.

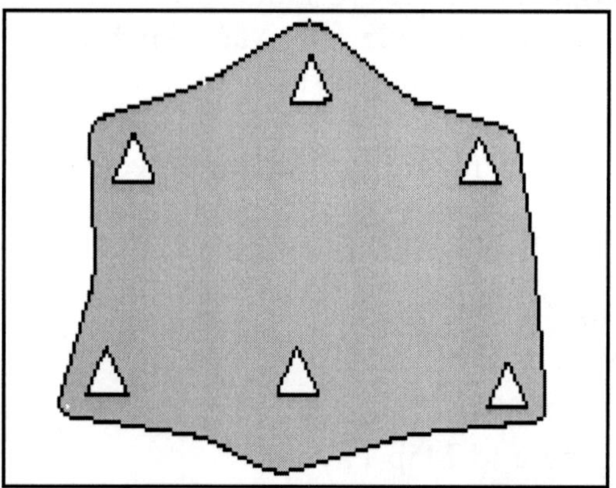

Fig. 6.1: Determination of Transformations

To do transformation, coordinates must be available in both coordinate systems. That is in WGS 94 and in the local system. At least three, preferably four common points should have been observed. The more common points included the more will be the precision. Common points are point for which coordinates and Orthometric heights are known in the local system. It is important to note that the transformation will apply to points in the area bound by the common points (Fig. 6.1). Points outside of this area should not be transformed using the calculated parameters but should form part of a new transformation area. It is common for GPS surveys to be measured to a much higher accuracy than older surveys measured with traditional optical instruments. In the vast majority of cases, the previously measured points will not be as accurate as the new points measured with GPS. This may create some time non-homogeneity in the network.

In order to transform a coordinate from one system to another, the origins and axes of the ellipsoid must be known relative to each other. From this information, the shift in space in X, Y and Z from one origin to the other can be determined. The following are a few important transformation approaches:

a. *Three-parameter methods:* The simplest datum transformation method is a geocentric, or three-parameter transformation. In the geocentric transformation, models the differences between two datums in the X,Y,Z coordinate system. One datum is defined with its center at 0,0,0. The center of the other datum is defined at some distance (DX,DY,DZ) in meters away.

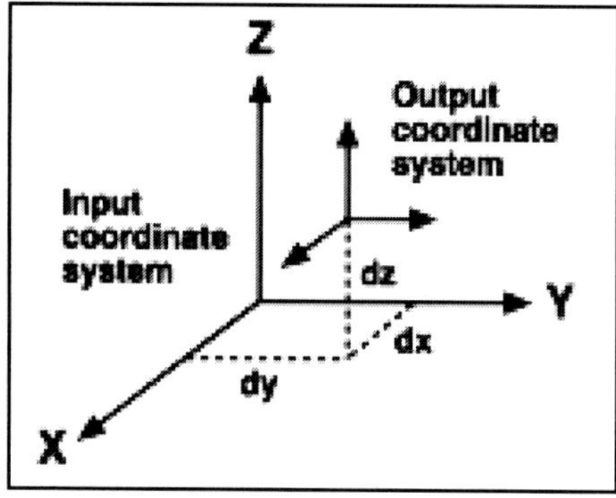

Fig. 6.2: Three-parameter Method

Usually the transformation parameters are defined as going "from" a local datum "to" WGS 1984 or another geocentric datum. The three parameters are linear shifts and are always in meters.

$$\begin{bmatrix} X \\ Y \\ Z \end{bmatrix}_{new} = \begin{bmatrix} \Delta X \\ \Delta Y \\ \Delta Z \end{bmatrix} + \begin{bmatrix} X \\ Y \\ Z \end{bmatrix}_{original}$$

Fig. 6.3: Three-parameter Method

b. *Seven-parameter methods:-* A more complex and accurate datum transformation is possible by adding four more parameters to a geocentric transformation. The seven parameters are three linear shifts (DX,DY,DZ), three angular rotations around each axis (rx,ry,rz), and scale factor(s).

$$\begin{bmatrix} X \\ Y \\ Z \end{bmatrix}_{new} = \begin{bmatrix} \Delta X \\ \Delta Y \\ \Delta Z \end{bmatrix} + (1+s) \cdot \begin{bmatrix} 1 & r_z & -r_y \\ -r_z & 1 & r_x \\ r_y & -r_x & 1 \end{bmatrix} \cdot \begin{bmatrix} X \\ Y \\ Z \end{bmatrix}_{original}$$

Fig. 6.4: Seven-parameter Method

The rotation values are given in decimal seconds, while the scale factor is in parts per million (ppm). The rotation values are defined in two different ways. It's possible to define the rotation angles as positive either clockwise or counterclockwise as you look toward the origin of the X,Y,Z systems.

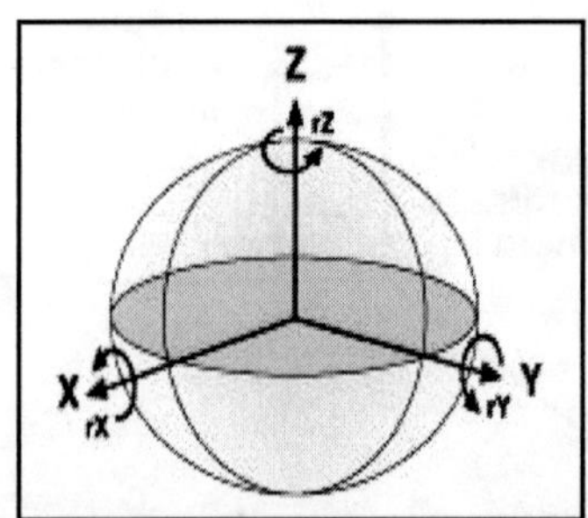

Fig. 6.5: Angular Rotations of Axis

The previous equation is called the Coordinate Frame Rotation transformation. The rotations are positive counterclockwise. Europe uses a different convention called the Position Vector transformation. Both methods are sometimes referred to as the Bursa–Wolf method.

c. *Molodensky Method:-* The Molodensky-Badekas method is a variation of the seven-parameter methods. It has an additional three parameters that define the XYZ origin of rotation. Sometimes this point is known as the origin of the datum, or geographic coordinate system. Given the XYZ origin of rotation point, it is possible to calculate an equivalent Coordinate Frame transformation. The DX, DY, and DZ values will change but the rotation and scale values will remain the same.

The Molodensky method converts directly between two geographic coordinate systems without actually converting to an X,Y,Z system. The Molodensky method requires three shifts (DX,DY,DZ) and the differences between the semi major axes (Da) and the flattening (Df) of the two spheroids.

$$
\begin{aligned}
(M+h)\Delta\varphi = & -\sin\varphi\cos\lambda\Delta X - \sin\varphi\sin\lambda\Delta Y \\
& + \cos\varphi\Delta Z + \frac{e^2\sin\varphi\cos\varphi}{(1-e^2\sin^2\varphi)^{1/2}}\Delta a \\
& + \sin\varphi\cos\varphi\left(M\frac{a}{b} + N\frac{b}{a}\right)\Delta f
\end{aligned}
$$

$$
(N+h)\cos\varphi\Delta\lambda = -\sin\lambda\Delta X + \cos\lambda\Delta Y
$$

$$
\begin{aligned}
\Delta h = & \cos\varphi\cos\lambda\,\Delta X + \cos\varphi\sin\lambda\,\Delta Y \\
& + \sin\varphi\,\Delta Z - (1-e^2\sin^2\varphi)^{1/2}\Delta a \\
& + \frac{a(1-f)}{(1-e^2\sin^2\varphi)^{1/2}}\sin^2\varphi\,\Delta f
\end{aligned}
$$

h ellipsoid height (meters)
j latitude
l longitude
a semi major axis of the spheroid (meters)
b semi minor axis of the spheroid (meters)
f flattening of the spheroid
e eccentricity of the spheroid

M and N are the meridional and prime vertical radii of curvature, respectively, at given latitude. The equations for M and N are:

$$M = \frac{a(1-e^2)}{(1-e^2\sin^2\varphi)^{3/2}}$$

$$N = \frac{a}{(1-e^2\sin^2\varphi)^{1/2}}$$

d. *Abridged Molodensky method:* - The Abridged Molodensky method is a simplified version of the Molodensky method. The equations are:

$$M\Delta\varphi = -\sin\varphi\cos\lambda\Delta X - \sin\varphi\sin\lambda\Delta Y + \cos\varphi\Delta Z + (a\Delta f + f\Delta a)\cdot 2\sin\varphi\cos\varphi$$

$$N\cos\varphi\Delta\lambda = -\sin\lambda\Delta X + \cos\lambda\Delta Y$$

$$\Delta h = \cos\varphi\cos\lambda\Delta X + \cos\varphi\sin\lambda\Delta Y + \sin\varphi\Delta Z + (a\Delta f + f\Delta a)\sin^2\varphi - \Delta a$$

e. *Grid-based transformation methods:-* Grid-based methods allow modeling the differences between the systems and are potentially the most accurate method. The area of interest is divided into cells. The National Geodetic Survey (NGS) publishes grids to convert between NAD 1927 and other older geographic coordinate systems and NAD 1983. Accuracies can vary depending on how good the geodetic data in the area was when the grids were computed.

6.4.3 Projection

Maps are flat, but the surfaces they represent are curved. Transforming three-dimensional space onto a two-dimensional map is called projection. Projection formulas are mathematical expressions that convert data from a geographical location— latitude and longitude—on a sphere or spheroid to a representative location on a flat surface.

Map projections are systematic transformations of the spheroidal shape of the Earth so that the curved, three-dimensional shape of a geographic area on the Earth can be represented in two dimensions, as x,y coordinates.

A projected coordinate system is defined as a flat, two-dimensional surface. Unlike a geographic coordinate system, a projected coordinate system has constant lengths, angles, and areas across the two dimensions. A projected coordinate system is always based on a geographic coordinate system that is based on a sphere or spheroid.

In a projected coordinate system, locations are identified by x,y coordinates on a grid, with the origin at the center of the grid. Each position has two values that reference it to that central location. One specifies its horizontal position and the other its vertical position. The two values are called the x-coordinate and y-coordinate. Using this notation, the coordinates at the origin are x = 0 and y = 0.

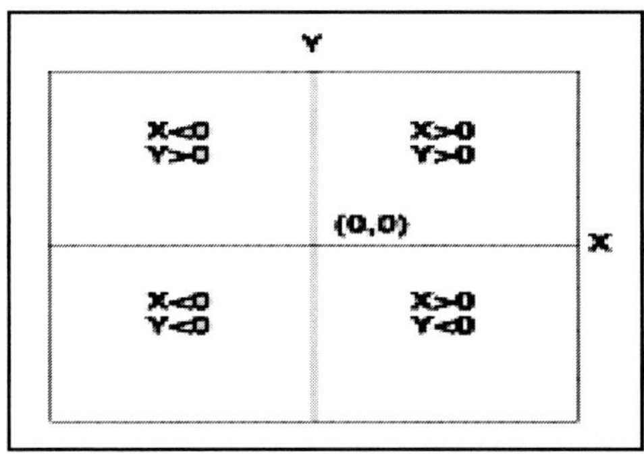

Fig. 6.6: The Logic of Projected Co-Ordinate System

On a gridded network of equally spaced horizontal and vertical lines, the horizontal line in the center is called the x-axis and the central vertical line is called the y-axis. Units are consistent and equally spaced across the full range of x and y. Horizontal lines above the origin and vertical lines to the right of the origin have positive values; those below or to the left have negative values. The four quadrants represent the four possible combinations of positive and negative x- and y-coordinates.

When working with data in a geographic coordinate system, it is sometimes useful to equate the longitude values with the X axis and the latitude values with the Y axis.

6.4.4 Mapping and Map Layout

Cartography can be described as the graphic principles supporting the art, science, and techniques used in making maps or charts. It was developed in a time before the computer and GIS. Traditionally, maps have been created to serve two main functions. The first function has been to store information. Creating a map has been a way to record information for future reference. The second function has been to provide a picture to relay spatial information to a user.

The purpose for designing a map is critical to its design. When designing a map, a map maker needs to know the answers to some fundamental questions, such as: What is being mapped? Who is the audience? How is this map being presented, on its own or as part of a report? What medium will be used to display this map?

Map formats: Generally maps are considered to be in two formats. One is a general reference map such as a USGS topographic map or the map of a city. In this form the map is providing information to convey where things are in relation to each other. The second is a thematic map, where the map is used to convey information about a particular theme or multiple themes, such as land use, population, or health statistics.

Basic mapping principles: There are many kinds of maps, each with general and possibly specific requirements. While a skilled cartographer is usually required to make maps with specific or special requirements, anyone can make good general and informative maps by considering the following simple guidelines. These guidelines have been organized into seven areas that you can use as a checklist for creating or improving your maps.

- Purpose—Typically, a map does not have more than one purpose. Trying to communicate too much in one map—having more than one purpose for the map—tends to blur the message and confuse the map reader. Using two or more maps, each focused on a single message, is always a better strategy.
- Audience—who will be reading your map? Are you designing a map for a few readers or for a large audience of hundreds or millions of people? It's better to target your map to the person least prepared to understand your map's message.
- Size, scale, and media—The physical size of a map relative to the geographic extent shown on the map will dictate the scale of the map and determine how you will represent the actual size and number of features shown on the map. Data is often collected at a particular scale. However, if you're not displaying the data at that scale, be sure your data "fits". For example, roads typically collected for 1:24,000 mapping will be far more detailed than needed for a smaller scale map (such as 1:2,000,000), so be sure to reduce the number of roads drawn on your map. Media also plays an import role, because a map printed on newspaper will not show fine details clearly, whereas one printed on high-quality paper will. In addition, the details on a digital map could vary depending on the viewing program. For example, a static map used in a Web page would be designed to encompass less information than one designed for browsing using a program such as ArcReader.

- Focus—Refers to where the designer wishes the map reader to first focus. Typically cool colors (blues, greens, and light gray) are used for background information, and warm colors (red, yellow, black) are used to capture the reader's attention.
- Integrity—you may want to cross-validate some of your information, such as the names or spelling of some features. If the data was produced by another organization, it is customary to give that organization credit on the map.
- Balance—how does your map look on the page or screen? Are the parts of the map properly aligned? The body of the map should be the dominant element. Try to avoid large open spaces. Be flexible in where you place elements (that is, not all titles need to go at the top). Should some components on the map be contained within a border?
- Completeness—a map generally should contain some basic elements, such as a title, legend, scale bar, and North arrow; however, there are exceptions. For example, if a graticule exists, it is not necessary to place a North arrow. Basically, place all the information you think your readers need to fully understand the map. Examples of the basic elements to include on a map are shown in the diagram titled 'Map elements diagram'.

Before publishing your map, it is always a good idea to have someone else look it over, especially for spelling and overall appearance.

6.5 Map Element

- Map body—the primary mapped area. You can display more than one image of your primary mapped area within your document. For example, you may want to portray change by showing several images with differing but related information, such as population maps of various years. Your map may also contain a locator map (a smaller-scale map used to help the reader in understanding where the main area of interest is located), an inset map (used to give more detailed information of an area within the main map that may not easily be understood), or an index map (often used to show where in a series of maps one map exists). All are used to assist in communicating your information to others. In ArcMap, each of these mapped areas is referred to as a data frame.

- Title—is used to tell the reader what the map represents. This is often placed on a map layout as text.
- Legend—lists the symbology used within the map and what it represents. This can be created using the Legend Wizard in the layout and edited further once created.
- Scale—Provides readers with the information they need to determine distance. A map scale is a ratio, where one unit on the map represents some multiple of that value in the real world. It can be numeric (1:10,000), graphic (a scale bar), or verbal (one inch equals 10,000 inches). Maps are often referred to as "large" or "small" scale. This "size" reference refers to the ratio (or fraction). For example, a 1:100 scale map is larger than a 1:10,000 map, because 1/100 (0.01) is a larger value than 1/10,000 (0.00001). A smaller scale map displays a larger area, but with less detail. The scale is inserted in the map layout view.
- Projection—is a mathematical formula that transforms feature locations from the Earth's curved surface to a map's flat surface. Projections can cause distortions in distance, area, shape, and direction; no projection can avoid some distortion. Therefore, the projection type is often placed on the map to help readers determine the accuracy of the measurement information they get from the map.

- Direction – this is shown using a North arrow. A map may show true north and magnetic north. This element is inserted in the map layout view.
- Data source – the bibliographic information for the data used to develop the map.

Other map components include (but are not limited to) dates, pictures, graticules or grids, reports, tables, additional text, neat lines, and authorship.

Chapter - 7

GPS APPLICATIONS

GPS receivers are an excellent navigation tool, with seemingly endless applications. GPS technology has enabled a revolution in the ways we move people, goods and information; build communities; manage the environment; predict the weather and natural disasters; and respond to emergencies. Combined with other Geomatics technologies, GPS data can be used in a wide range of applications, including locating and tracking vehicles and other objects, managing infrastructures, time-stamping information and images, and navigating between points on the globe.

7.1 Five Main Uses of GPS

GPS technology has matured into a resource that goes far beyond its original design goals. These days' people from a variety of professions are using GPS in ways that make their work more productive, safer, and sometimes even easier. There are five main uses of GPS today:

Location- determining a basic position.

Navigation - getting from one location to another.

Tracking - monitoring the movement of people and things.

Mapping- creating maps.

Timing - providing precise timing.

Other uses – Precision agriculture, mobile mapping, etc.

7.1.1 Location

The first and most obvious application of any Location Based Service such as GPS is the simple determination of a "position" or location. GPS was the first positioning system to offer highly precise location data for any point on the planet, in any weather. Knowing the precise location of something, or someone, is especially critical when the consequences of inaccurate data are measured in human terms.

7.1.2 Navigation

GPS helps you determine exactly where you are, but sometimes it is more necessary to know how to get somewhere else. GPS was originally designed to provide navigation information for ships and planes. So it's no surprise that while this technology is appropriate for navigating on water, it's also very useful in the air and on the land.

7.1.3 Tracking

GPS used in conjunction with communication links and computers can provide the backbone for systems tailored to applications in agriculture, mass transit, urban delivery, public safety, and vessel and vehicle tracking. Therefore, more and more police, ambulance, and fire departments are adopting systems like GPS-based AVL (Automatic Vehicle Location) Manager to pinpoint both the location of the emergency and the location of the nearest response vehicle on a computer map. With this kind of clear visual picture of the situation, dispatchers can react immediately and effectively.

7.1.4 Mapping and Surveying

Mapping the planet has never been an easy task, but GPS today is being used to survey and map it precisely, saving time and money in this most stringent of all applications. GPS can help generate maps and models of everything in the world, mountains, sea, rivers, and cities and help manage endangered animals, archaeological treasures, precious minerals and all sorts of resources, as well as accurately managing the effect of damage and disasters.

For example peace brokers used GPS generated maps to determine the partitions of Bosnia under the Dayton Peace Accord, and GPS was used to map Cambodia accurately for the UN stabilization force introduced after the civil war ended in 1993. In fact large tracks of Cambodia had never been mapped at all, due to inhospitable terrain and low population density. GPS solved this in weeks, whereas it would have taken months prior to the introduction of GPS.

7.1.5 Timing

GPS can also be used to determine precise time, time intervals, and frequency. There are three fundamental ways we use time:

As a universal marker,

As a way to synchronize, and

To provide an accurate, unambiguous sense of duration.

As discussed before, GPS satellites carry highly accurate atomic clocks. And in order for the system to work, GPS receivers here on the ground synchronize themselves to these clocks. That means that every GPS receiver is, in essence, an atomic accuracy clock. Astronomers, power companies, computer networks, communications systems, banks, and radio and television stations can benefit from this precise timing.

A standard GPS receiver will not only place us on a map at any particular location, but will also trace our path across a map as we move. It can tell us:

- How far we have traveled (odometer);
- How long we have been traveling;
- Our current speed (speedometer);

- Our average speed;
- A "breadcrumb" trail showing you exactly where you have traveled on the map; and
- The estimated time of arrival at our destination if we maintained our current speed.

Most receivers have a certain amount of memory available for users to store their own navigation data. This greatly expands the functionality of the receiver, because it essentially lets its user make a record of specific points on Earth. The basic unit of user input is the *waypoint*. A waypoint is simply the co-ordinates for a particular location. We can save this in our receiver's memory in two ways:

- We can tell the receiver to record its co-ordinates when we are at that location; or
- We can find the location on a map (the internal map or another one) and enter its co-ordinates as a waypoint.

This capability lets us use our GPS receiver in a number of different ways. We can record any specific location that interests us, so that we will be able to find it again at a later time, e.g., record where we left our car. We can also combine a series of different waypoints to form a route. One way to use this function is to periodically record waypoints as we make a trip so that we can backtrack, or follow the same route again on another trip. Route mapping also lets us plan ahead. When we have time to examine a map at home, we can record a series of waypoints along the roads or trails that lead to our destination. Then, when we are traveling, all we will need to find our way is our GPS receiver. As we travel, the receiver will show us which way to go and give us the distance to our next waypoint. All we need to do is follow its simple directions!

Receivers with route capabilities will let us save a certain number of waypoints to memory so that we can use them again and again. If the receiver has a data port, we can also download our routes to a computer, which has much more storage memory, and then upload them again when we plan to follow those routes.

Because they have so much more storage capability, computers can do a lot more with GPS location data than our average receiver. A receiver with a data port can feed the raw location co-ordinates into a computer running more complicated software. There are a number of available software applications that can place us on detailed maps of particular areas. If we want to use our receiver for complicated navigation, down back-roads for example, this capability will help us out tremendously. We can also update our computer maps, so that they include any surveying adjustments or changes in an area, whereas a receiver's onboard map usually cannot be changed. When we use our receiver in conjunction with a computer, we increase the receiver's capabilities considerably. Also, our receiver will not be outdated as quickly, because in conjunction with a computer, all it needs to do is provide co-ordinates - the computer does the rest.

7.2 Applications in Transportation and Emergency Situations

GPS technologies are dramatically improving transportation on land, at sea and in the air. GPS has been called the most significant development in air navigation and control since radar – and with good reason. It has improved safety by allowing better management of the air corridors, and will reduce fuel consumption by helping establish more efficient routes and schedules. At sea, GPS also helps vessels reach their destinations safely and efficiently. But the largest transportation market for GPS lies in land navigation and vehicle positioning. GPS technology is used to dispatch police, ambulances and fire fighters in emergency situations, reducing the response time and saving lives. A vehicle-mounted GPS system has been developed that records road conditions and roadside features, maps transportation corridors and performs other useful functions. The technology is even being used to locate stolen vehicles.

7.3 Infrastructure Building, Agricultural and Environmental Applications

GPS equipment and software is being used to speed the infrastructure building process and to provide precise positions for the creation of databases of resources and the production of maps that can be used to evaluate the feasibility and impact of programs and projects and help implement them. For example, GPS can be used to improve fertilizing and harvesting operations by directing the movement of farming equipment. Images from remote sensing satellites provide information about an area's fertilization/harvesting needs, and GPS is then used to guide the application of the fertilizer or the harvesting equipment. GPS can be also used to support the investigation of hazardous waste sites, the mapping of ecosystems, the monitoring of oil spills and clean-up efforts, and the tracking and mapping of airborne pollutants.

7.4 GPS and Crop Production

The Global Positioning System (GPS) provides opportunities for agricultural producers to manage their land and crop production more precisely. Common names for general GPS applications in farming and ranching include precision agriculture, site-specific farming and prescription farming. GPS applications in farming include guidance of equipment such as sprayers, fertilizer applicators and tillage implements to reduce excess overlap and skips. They can also be used to precisely locate soil- sampling sites, map weed, disease and insect infestations in fields and apply variable rate crop inputs, and, in conjunction with yield monitors, record crop yields in fields.

7. 4.1 Equipment Guidance Systems

Light bar-guided and automated steering systems help maintain precise swath-to-swath widths. Guidance systems identify an imaginary A-B starting line, curve or circle for parallel swathing using GPS positions and a control module. The module takes into account the swath width of the implement and then uses GPS to guide machines along parallel, curved, or circular evenly spaced swaths. Guidance systems include a display module that uses audible tones or lights as directional indicators for the operator. The guidance system allows the operator to monitor the light bar to maintain the desired distance from the previous swath.

Guidance systems require two principle components: a light bar or screen, which is essentially an electronic display showing a machine's deviation from the intended position, and a GPS receiver for locating the position. This receiver must be designed for this purpose and it must operate at a higher frequency (position calculations are usually 5 to 10 times per second) than a GPS receiver designed to record positions for a yield monitor. GPS receivers designed for guidance can be used in conjunction with a yield monitor or for other positioning equipment. Automated steering systems integrate GPS guidance capabilities into the vehicle steering system. Automated steering frees the operator from steering the equipment except at corners and at the ends of fields. Guidance systems base prices are approximately $3,000, including the GPS receiver and a readout unit. Systems that steer the vehicle will be higher priced.

7.4.2 Yield Monitoring Systems

Yield monitoring systems typically utilize a mass flow sensor for continuous measuring of the harvested weight of the crop. The sensor is normally located at the top of the clean grain elevator. As the grain is conveyed into the grain tank, it strikes the sensor and the amount of force applied to the sensor represents the recorded yield. While this is happening, the grain is being tested for moisture to adjust the yield value accordingly. At the same time, a sensor is detecting header position to determine whether or not yield data should be recorded. Header width is normally entered manually into the Monitor and a GPS, radar or a wheel rotation sensor is used to determine travel speed. The data is displayed on a monitor located in the combine cab and stored on a computer card for transfer to an office computer for analysis. Yield monitors require regular calibration to account for varying conditions, crops and test weights.

7.4.3 Field Mapping with GPS and GIS

GPS technology is used to locate and map regions of fields such as high weed, disease and pest infestations. Rocks, potholes, power lines, tree rows, broken drain tile, poorly drained regions and other landmarks can also be recorded for future reference. GPS is used to locate and map soil-sampling locations, allowing growers to develop contour maps showing fertility variations throughout fields. The various datasets are added as map layers in geographic information system (GIS) computer programs. GIS programs are used to analyze and correlate information between GIS layers.

7.4.4 Precision Crop Input Applications

GPS technology is used to vary crop inputs throughout a field based on GIS maps or real-time sensing of crop conditions. Variable rate technology requires a GPS receiver, a computer controller, and a regulated drive mechanism mounted on the applicator. Crop input equipment such as planters or chemical applicators can be equipped to vary one or several products simultaneously. Variable rate technology is used to vary fertilizer, seed, herbicide, fungicide and insecticide rates and for adjusting irrigation applications.

7.5 GPS based Vehicle Tracking Systems

GPS is used for recording the movement of the vehicle. A GPS based unit, fitted inside the vehicle, will store the desired parameters - lat, log, speed, etc. at a predefined interval. It is important to note that - as tracking/recording is done by GPS based system - its coverage is 100 per cent and can trace every inch of the distance. And once the vehicle is at the control station, the data from the instrument is downloaded by software, on a regular PC, for analysis. The downloaded data contain the route, path that particular vehicle has traveled. One can generate different kinds of reports and GIS map plots with these data. Map plots make it easy to visualize the movement of the vehicle. And if the data are maintained in database with relation to truck and the driver, one can view the pattern of a particular truck or driver over a period of time. This kind of information can be very useful in taking decisions for future. This is an inexpensive, useful and important option for goods-tracking requirements in India.

In case one needs to track a vehicle in real time then online tracking is the best way of doing it. The 'In Vehicle' unit takes the current location/position information provided by GPS and transmits it to the base the station server via a communication link. Optionally it may also save the location information in unit's memory. As soon as the location information is received at the base station server, the software at the server saves the information, processes it and shows the current location on a GIS map - in real time.

7.6 Using GPS to Create and Update Maps

In Kenya, the Division of Parasitic Diseases of the Centres for Disease Control and Prevention (CDC, Atlanta, Georgia, US) works with the Kenya Medical Research Institute to study malaria and to work to prevent it. Nearly three hundred researchers work on various projects near Lake Victoria and Kisumu, Kenya's third largest town. These researchers use Differential Global Positioning Systems (DGPS) to collect positions and data in the field, and then edit and analyze this data in ArcView GIS. One study region had its last map made in the late 1960s, and researchers needed an updated map for their study. Researchers began by creating an updated map of a 225 square mile region just outside of Kisumu along Lake Victoria using GPS. The DGPS mapping team hired local fishermen to row them in small fishing boats to map the shore of the lake. Roads were mapped by driving cars along them while a team member captured location data with the DGPS. Once they had an updated map of the region, they could begin using their GIS and create the maps that would help them understand the impact of bednets (treated with odourless insecticides) on malaria, childhood mortality, and mosquito populations. (Lang, 2000).

7.7 Digital Angel

The idea behind the initial version of Digital Angel is to build a microchip that can be worn close to the body. This microchip will include bio-sensors that will measure the biological parameters of the body and store this information. It will also have an antenna that will receive signals from GPS satellites. The geographical location of the chip can be derived from these signals. The antenna also communicates with ground stations. It will receive commands from the stations and will send the biological information and location data to the ground station. This could take the form of a distress signal sent to a monitoring facility when the unit detects a medical emergency.

7.8 Other Applications

More and more producers today are using precision farming techniques that can help increase profits and protect the environment. Precision or site-specific farming involves applying fertilizer, pesticides and other inputs only where they are needed. GPS-guided equipment is often used for variable rate application of fertilizer (based on soil tests) or pesticides (based on pest survey). GPS can also be used to develop the initial reference maps upon which variable rate applications are based. A GPS system on a combine with a yield monitor can be used to develop an on-the-go yield map. Mounted in an airplane, GPS can be used to guide aerial spraying operations. GPS can be used to locate weed, insect or diseases infestations and monitor their spread. It can also be used to navigate back to previously mapped infestations to apply controls.

A field map can be created using GPS to record the coordinates of field borders, fence lines, canals, pipelines, and point locations such as wells, buildings, and landscape features. The resulting field map might be the first layer a producer would develop for an on-farm GIS. Additional layers showing crop damage from hail or drought, and riparian areas or wetlands could be mapped using GPS. Ranchers could use GPS to develop rangeland utilization maps and to navigate back to previously mapped areas or monitoring sites

Mobile computing and wireless communication are two of the strongest trends in the computer industry today. Mobile computing is bringing fundamental changes in the way geography is being utilized. It has the ability to connect one's work directly with the world around. Mobile computing comprises the integration of three merging technologies—lightweight hardware, global positioning system (GPS), and wireless communication.

GPS is being used for emergency response (fire, ambulance, and police), search and rescue, fleet management (trucking, delivery vehicles, and public transportation) and for automobile guidance systems. Recreational uses of GPS include navigation while hiking, hunting, or skiing. GPS is even used on golf courses to track golf carts, and to let players know how far it is to the center of the greens.

Risk assessment and hazard warning systems for marine navigation are being developed using GPS. In the air, GPS is being used for en-route navigation (helicopter, airplane, and hot-air balloon), aircraft landing, and air-collision avoidance systems.

The applications of GPS for natural resources professionals fall into two primary categories and are dependent upon the accuracy requirements of the data. The first category is primary data acquisition and usually includes some link to a geographic information system (GIS). It may or may not include integration with other data sources. The second category is navigation. The accuracy dependence of the data is generally met by the choice of receiver.

Primary data acquisition is generally of a mapping nature and is most efficient when collecting data not found or recognized on current aerial photography. Updating includes new roads constructed since the time of the photography is one example of this. Generally, it is not cost effective to do mapping work with GPS when areas of interest are readily mapable from aerial photos. However, GPS may be very effective when used to provide ground control coordinates for use in stereo plotting mapping applications.

Other uses of GPS include real estate valuation and taxation assessment, air quality studies, environmental protection, demographic analysis including marketing studies, atmospheric studies, oil and gas exploration, and scientific exploration. There are many additional current and possible uses for GPS. Any application where location information is needed is a possible candidate for GPS.

7.9 Theme wise Classification of Applications of GPS

7.9.1 Land Applications

- Surveying and mapping, including cadastral and urban networks, data capture surveys for Geographic Information Systems (GIS), engineering surveys, photogrammetrical control (airborne and terrestrial), and geophysical resource surveys.

- Geodetic applications, including the establishment of control networks over regional and continental extent, height and geoid determination, precise engineering and subsidence monitoring surveys.

- Geodynamic applications, for measuring the relative position of a regional network at regular intervals in order to infer horizontal and vertical crustal motion.

- Land navigation, to support emergency vehicles (police, search & rescue, etc.) and for the monitoring of cars, taxis, dangerous and valuable cargoes, trucks and railways.

- Transportation and communication, to support aids for navigation for land, sea and air users, land operations taking advantage of permanent GPS stations, time transfer operations, etc.
- Recreational uses, for hiking, orienteering, etc.

7.9.2 *Areal Applications*

- Airport approach and landing, as an aid for all categories of landing including full instrument landings.
- Domestic and intercontinental enroute navigation, including helicopter operations.
- Air Traffic Control operations, dynamic routing, new airport approaches.
- Search and rescue operations, including coordinated search operations, collision avoidance, and rendezvous.
- Aerial photogrammetry, including laser profiling and radar imaging.
- Airborne geophysical surveys, including position and attitude determination for gravimeter, magnetometer and remote sensing operations.
- Recreational applications, glider, ballooning, light aircraft, and parachuting operations.

7.9.3 *Marine Applications*

- Open ocean and coastal navigation, with or without aid of electronic charts.
- Harbour and inland waterway navigation, including when visibility is low.
- Search and rescue operations, including coordinated search operations and rendezvous.
- Ship Monitoring Systems for collision avoidance, remote piloting, and tracking of fishing fleets and dangerous cargoes.

- Offshore geophysical surveys, including seismic and gravity surveys.
- Engineering applications such as drill rig, pipeline and other offshore structure positioning.
- Hydrographic surveys, including surveys to support oceanographic research.
- Recreational applications, such as fishing, snorkelling, sailing.

7.9.4 Space Applications

- Spacecraft launch and landing.
- Navigation, in-flight, re-entry and rendezvous, for Earth-orbiting and interplanetary missions.
- Orbit determination, for many Earth resource and scientific missions.
- Sounding the atmosphere, including GPS "meteorology", and ionospheric studies.

7.9.5 Military Applications

- Enroute and low-level navigation for tactical and strategic operations.
- Target acquisition, including forward observation, covert operations.
- Photo reconnaissance and intelligence gathering.
- Remotely operated vehicles, for reconnaissance and targeting.
- Weapon guidance and control, "smart" bombs, cruise missiles, etc.
- Command and control, the C^3 applications, tracking of battle elements.
- Updating inertial navigation systems, at sea, in the air and on land.
- Fleet, air and land operations, the "electronic battlefield", battle element maneuvering.

7.10 Trends in GPS Applications

- There will be a blurring of the distinctions between GPS navigation, surveying and geodesy.
- The number of civilian users will continue to grow, compared to the number of military users - the ratio is currently 9:1 and will increase.
- GPS will facilitate the "position information society" - where location will be a common and useful piece of information for everyday activities.
- "Consumer-based" GPS products will make up approximately 50 per cent of the market within the next few years.

7.10.1 GPS Navigation

- Expanding range of applications as the cost of hardware drops.
- GPS will be the primary navigation tool for air, land and sea travel.
- Largest market will be for land vehicles, for ITS applications - many applications, from autonomous navigation to "fleet management".
- GPS will likely become an insignificant component of larger and more complex system - e.g. "fleet monitoring", "Intelligent Transportation System", "Future Air Navigation System", etc.
- The elimination of Selective Availability will make possible absolute positioning accuracy at the 10-20 metre level.
- DGPS use will expand however, requiring the establishment of transmitting base stations. GPS and Map Display technology will converge into new navigation products - e.g. ITS, Electronic Chart Systems, etc.
- Satellite communications will play a large role in GPS and DGPS applications - particularly for sea and air applications.

- Wide-Area DGPS will be refined, and encroach on Local-Area DGPS.
- Increasing use of low-cost handheld receivers used for geo-coding (data capture) in support of GIS applications.
- Increasing integration of GPS with other navigation sensors - e.g. GPS and Dead Reckoning systems.

7.10.2 GPS Surveying

- GPS will be a commonly used technique for all surveys, especially for distances greater than about 5km.
- GPS surveying applications will expand into the engineering and cadastral areas - changes in Surveyors' law may be required.
- Increased use of "non-conventional" positioning techniques - e.g. new methodologies using GPS survey receivers, but also increased use of low-cost handheld receivers for sub metre accuracy applications.
- Base station operation will be increasingly supported - post-processing as well as real-time mode.
- Increasing use of precise single point positioning techniques based on IGS orbit and satellite clock correction products.
- The World Wide Web will play an ever increasing role for GPS "transactions", including access to base station data, baseline reduction services, etc.
- GPS deformation surveys - engineering structures, Earthquake and volcanic zones, etc.
- Mixing of GPS survey receivers will no longer be unusual.
- More and more applications will be addressed in real-time.
- Precision airborne GPS will be used extensively for remote sensing, airborne geophysics and photogrammetry.

- Refinement of one-epoch carrier phase positioning - no longer a distinction between static and kinematic surveying.

7.10.3 GPS Leveling

- A viable tool for 3rd or 4th order leveling.
- A viable tool for checking conventional leveling networks.
- Increased range of geoid computation packages and/ or geoid map products will be available.
- Research effort directed towards attaining 0.01ppm height accuracies in GPS geodesy - for monitoring sea level rise, subsidence of land, etc.

7.10.4 GPS Geodesy

- Accuracies at the few 0.01ppm level will become routine.
- GPS geodesy will play the dominant role in geodynamic studies.
- GPS will be used for other non-positioning applications such as ionosphere monitoring and the determination of tropospheric conditions - GPS "meteorology"!
- GPS geodesy will be performed using permanent GPS networks, as well as the traditional "campaign-style" survey.
- GPS geodesy will play the dominant role in geodynamic studies.
- The IGS will support all precise GPS surveys.
- Improvements in instrumentation, reference system definition (ITRS), global tracking (IGS), observation modeling and network design.
- All of these will require greater partnership between academics, researchers, government survey organizations, instrument manufacturers - GPS geodesy will be less the prerogative of a select band of

experts, and the government agencies will take on more of the role.

- More frequent GPS campaigns for geodynamic studies, at a lower cost and on a local area basis - with IGS there is no need for orbit computation!
- Software refinements to make data reduction easier - data processing are the biggest bottleneck at present.
- Permanent arrays or networks of precision GPS receivers operating 24hr/day - on global, continental, national and local scales for different applications.

7.11 Misuses of GPS

Both the NAVSTAR and GLONASS systems were conceived as military systems, intended to augment military weaponry in times of war. However, GPS is a classic example of a dual use system. That is, one which has both military and civilian uses. In this case, GPS is an extremely expensive system which has potentially massive civilian benefits.

7.11.1 Situations in Which Misuse Could Occur

7.11.1.1 Attacker vs. Country of Origin

In this case, the enemy would be directly attacking the GPS system's creator by using its own system against it. Example: Iraq used NAVSTAR augmented weapons vs. the United States.

7.11.1.2 Attacker vs. Ally

In this case, the enemy would be attacking an ally of the GPS system's creator used the GPS system. Example: Iraq used NAVSTAR augmented weapons vs. Saudi Arabia.

7.11.1.3 Attacker vs. Other Nation

In this case, the attacker would be attacking a nation not allied with the GPS system's creator. Example: Libya used NAVSTAR augmented weapons vs. Chad.

7.11.2 Types of Misuse

7.11.2.1 Intelligence Collection

GPS combined with laser range finding systems permits accurate positioning of potential targets from a distance. Information resulting from such data collection can be attacked at a later time using a variety of methods. These would include infantry assault, air attack, artillery, etc.

7.11.2.2 Augmentation of Existing Weapon Systems

Potential for terrorist use in short range mortar attacks. Post processed mapping grade differential GPS combined with commercial laser rangefinders could be used to surreptitiously map potential targets with respect to potential hiding spots for mortars. This would allow for better first shot accuracy with most any indirect fire weapon.

Similarly, military artillery batteries can shorten the time needed to survey in guns before they begin operation. This is relatively important in modern warfare because artillery batteries must move often to keep from being hit from counter fire. This principle is demonstrated in the U.S. Army's MLRS surface to surface missile system. The MLRS vehicle can carry up to eight unguided rockets in one tracked vehicle. An inertial guidance system in the vehicle is used to position the vehicle and aim the launch box at the target. A single MLRS vehicle, such as the one shown below, can stop, aim, shoot and leave within four minutes while providing the firepower of an entire battery.

7.11.2.3 Integration into Ballistic Missile Systems

A study has shown that the accuracy of short range ballistic missiles such as the Scud or the No Dong I can be improved by 20 to 25 per cent by using the civilian SPS GPS signals. The GPS receiver is used to provide more accurate boost phase velocity measurements to the guidance system. Differential correction provides little added benefit because other systemic errors begin to dominate. Longer range ballistic missiles benefit more because they are more affected by boost velocity errors.

7.11.2.4 Integration into Cruise Missile Systems

Biological weapons can be dispersed more easily and more effectively from a cruise missile than from a ballistic missile. Even civilian grade GPS provides sufficient accuracy for a low cost cruise missile to lay down a biological toxin such as anthrax in a desired pattern.

Chapter - 8

COMPETITORS TO GPS

At present GPS is the only effective satellite system which provides location based positioning services. However there are other positioning systems are also available, while some other are in the developmental stage, they are as follows:

8.1 GLONASS

The Russian government has developed a system, similar to GPS, called GLONASS. The first GLONASS satellite launch was in October 1982. The full constellation consists of 24 satellites in 3 orbit planes, which have a 64.8 degree inclination to the Earth's equator. The GLONASS system now consists of 12 healthy satellites. It is easy to presume that GLONASS is the same as GPS. Rather, there are some key differences between the two systems. GLONASS uses the same code for each satellite and many frequencies, whereas GPS which uses two frequencies and a different code for each satellite. Some GPS receiver manufacturers have incorporated the capability to receive both GPS and GLONASS signals. This increases the availability of satellites and the integrity of combined system.

In many ways, GLONASS is a more elegant and economical system design than GPS. Unfortunately there are a number of quality control issues and much more serious funding problems. The GLONASS system was declared fully operational on 18 January 1996. Since then there have been few occasions when the full complement of 24 satellites was operating. In addition to satellite outages eight satellites have been withdrawn from service since this date. So by late 1997 only 15 satellites were available for navigation.

Recent announcements by Russian officials may mean that 6 years of neglect of the GLONASS constellation of satellites will stop. Solid government financial commitments have been obtained to return the system to fully operational status by 2006, Russian officials say. Also, this new generation of GLONASS satellites (Glonass-M) will have longer in-orbit lives. They are designed to have a seven-year service life, compared to three years for the current GLONASS. Whether this financial commitment will materialize is another thing, but even in its current weakened state, GLONASS still has potential as a stand-alone navigation system and as an augmentation to GPS.

8.2 GALILEO

Galileo is Europe's contribution to the next generation Global Navigation Satellite System (GNSS). Unlike GPS, which is funded by the public sector and operated by the U.S. Air Force, Galileo will be a civil-controlled system that draws on both public and private sectors for funding. The service will be free at the point of use, but a range of chargeable services with additional features will also be offered. It is projected by 2008 a constellation of 30 satellites should be available. The technology behind Galileo is designed to be more accurate and more reliable than GPS or GLONASS. This will allow safety-critical systems - such as air traffic control, and ship and car navigation - to be run on the technology. The system should also guarantee coverage to previously inaccessible areas such as those that are either blocked by buildings or isolated areas at high latitudes.

However, it is estimated that the Galileo project could cost more than €3bn ($2.6bn). With some of the member countries facing recession, now might not be the time to fund the idea. Britain and the Netherlands have already questioned the timing, and Germany raised issues over costs. In contrast, France and Italy whose aerospace industries both stand to benefit from the system are looking forward to a go-ahead. The decision to push ahead with the project remains up in the air for now.

APPENDIX - I

ATOMIC CLOCKS

GPS satellites use atomic clocks, precise lock that depends for its operation on an electrical oscillator regulated by the natural vibration frequencies of an atomic system e.g. as a beam of cesium atoms. Time is measured by the movement of electrons in cesium atoms. They even keep time better than the rotation of the Earth and the movement of the stars. Atomic clocks are not radioactive. They do not rely on atomic decay: Radioactive decay is a natural process. The big difference between a standard clock in home and an atomic clock is that the oscillation in an atomic clock is between the nucleus of an atom and the surrounding electrons. The oscillation frequencies within the atom are determined by the mass of the nucleus and the gravity and electrostatic spring between the positive charge on the nucleus and the electron cloud surrounding it.

Types of Atomic Clocks

Though there are different types of atomic clocks, the principle behind all of them remains the same. The major difference is associated with the element used and the means of detecting when the energy level changes. The various types of atomic clocks include:

- Cesium atomic clocks employ a beam of cesium atoms (*Cesium 133* - An isotope of cesium). The clock separates cesium atoms of different energy levels by magnetic field.
- Hydrogen Atomic Clocks maintain hydrogen atoms at the required energy level in a container with walls

of a special material so that the atoms don't lose their higher energy state too quickly.

- Rubidium Atomic Clocks, the simplest and most compact of all, use a glass cell of rubidium gas that changes its absorption of light at the optical rubidium frequency when the surrounding microwave frequency is just right.

The most accurate atomic clocks available today use the cesium atom and the normal magnetic fields and detectors. In addition, the cesium atoms are stopped from zipping back and forth by laser beams, reducing small changes in frequency due to the Doppler Effect.

Working Principles of Cesium Clock

Atoms have characteristic oscillation frequencies. An atom will have many frequencies, some at radio wavelength, some in the visible spectrum, and some in between the two. Cesium 133 is the element most commonly chosen for atomic clocks. To turn the cesium atomic resonance into an atomic clock, it is necessary to measure one of its transition or resonant frequencies accurately. This is normally done by locking a crystal oscillator to the principal microwave resonance of the cesium atom. This signal is in the microwave range of the radio spectrum, and just happens to be at the same sort of frequency as direct broadcast satellite signals.

Atomic Clocks and GPS

Each of the Block II/IIA satellites carries two cesium (CS) and two rubidium (RB) atomic clocks. For Block II/IIA, CS clocks are considered the best satellite clocks. One of the atomic clocks defines space vehicle time and others operate as spare. The satellite clock errors are estimated by the IGS in connection with their satellite ephemeris production using a global data set.

APPENDIX - II

Leap Seconds

INTERNATIONAL EARTH ROTATION AND REFERENCE SYSTEMS SERVICE (IERS).

SERVICE DE LA ROTATION TERRESTRE

OBSERVATOIRE DE PARIS

61, Av. de l'Observatoire 75014 PARIS (France)

Tel. : 33 (0) 1 40 51 22 26

FAX : 33 (0) 1 40 51 22 91

e-mail : services.iers@obspm.fr

http://hpiers.obspm.fr/eop-pc

Paris, 4 July 2005

Bulletin C 30

To authorities responsible for the measurement and distribution of time.

UTC TIME STEP

on the 1st of January 2006

A positive leap second will be introduced at the end of December 2005.

The sequence of dates of the UTC second markers will be:

2005 December 31, 23h 59m 59s

2005 December 31, 23h 59m 60s

2006 January 1, 0h 0m 0s

The difference between UTC and the International Atomic Time TAI is:

from 1999 January 1, 0h UTC, to 2006 January 1 0h UTC:

UTC-TAI = - 32s

from 2006 January 1, 0h UTC, until further notice :

UTC-TAI = - 33s

Leap seconds can be introduced in UTC at the end of the months of December or June, depending on the evolution of UT1-TAI. Bulletin C is mailed every six months, either to announce a time step in UTC or to confirm that there will be no time step at the next possible date.

APPENDIX - III

International GPS Service

Since the late 1980s, the U.S. Global Positioning System (GPS) constellation of satellites has come to play a major role in regional and global studies of Earth. In the face of continued growth and diversification of GPS applications, the worldwide scientific community has made an effort to promote international standards for GPS data acquisition and analysis, and to deploy and operate a common, comprehensive global tracking system. As part of this effort, the IGS was formally recognized in 1993 by the International Association of Geodesy (IAG), and began routine operations on January 1, 1994, providing GPS orbits, tracking data, and other data products in support of geodetic and geophysical research. In particular, since January 1994, the IGS has made available to its user community the IGS official orbit based on contributions from the seven current IGS Analysis Centers. The IGS also supports a variety of governmental and commercial activities and develops international GPS data standards and specifications.

The Mission

The primary mission of the International GPS Service, as stated in the organization's 2002-2007 Strategic Plan, is: "The International GPS Service is committed to providing the highest quality data and products as the standard for global navigation satellite systems (GNSS) in support of Earth science research, multidisciplinary applications, and education. These activities aim to advance scientific understanding of the Earth system components and their interactions, as well as to facilitate other applications benefiting society."

To accomplish its mission, the IGS has a number of components: an international network of over 350 continuously operating dual-frequency GPS stations, more than a dozen regional and operational data centers, three global data centers, seven analysis centers and a number of associate or regional analysis centers. The Central Bureau for the service is located at the Jet Propulsion Laboratory, which maintains the Central Bureau Information System (CBIS) and ensures access to IGS products and information. An international Governing Board oversees all aspects of the IGS. The IGS is an approved service of the International Association of Geodesy since 1994 and is recognized as a member of the Federation of Astronomical and Geophysical Data Analysis Services (FAGS) since 1996.

The Operations

The IGS has developed a worldwide system comprising satellite tracking stations, Data Centers, and Analysis Centers to put high-quality GPS data and data products on line within a day of observations. For example, following the Northridge, California, Earthquake in January 1994, analysis teams using IGS-supplied data and products were able to quickly evaluate the disaster's immediate effects by determining station displacements accurately to within a few millimeters.

The IGS global network of permanent tracking stations, each equipped with a GPS receiver, generates raw orbit and tracking data. The Operational Data Centers, which are in direct contact with the tracking sites, collect the raw receiver data and format them according to a common standard, using a data format called Receiver Independent Exchange (RINEX). The formatted data are then forwarded to the Regional or Global Data Centers. To reduce electronic network traffic, the Regional Data Centers are used to collect data from several Operational Data Centers before transmitting them to the Global Data Centers. Data not used for global analyses are archived and available for online access at the Regional Data Centers. The Global Data Centers archive and provide on-line access to tracking data and data products.

Products and Applications

The IGS collects, archives, and distributes GPS observation data sets of sufficient accuracy to meet the objectives of a wide range of scientific and engineering applications and studies. These data sets are used to generate the following products:

GPS satellite ephemerides

GLONASS satellite ephemerides

Earth rotation parameters

IGS tracking station coordinates and velocities

GPS satellite and IGS tracking station clock information

Zenith tropospheric path delay estimates

Global Ionospheric Maps

IGS products support scientific activities such as improving and extending the International Earth Rotation Service (IERS) Terrestrial Reference Frame (ITRF), monitoring deformations of the solid Earth and variations in the liquid Earth (sea level, ice sheets, etc.), and in Earth rotation, determining orbits of scientific satellites and monitoring the ionosphere. For example, geodynamics investigators who use GPS in local regions can include data from one or more nearby IGS stations, fix the site coordinates from such stations to their ITRF values, and more importantly, use the precise IGS orbits without further refinement. Data from an investigator's local network can then be analyzed with maximum accuracy and minimum computational burden. Furthermore, the results will be in a well-defined global reference frame.

An additional aspect of IGS products is for the densification of the ITRF at a more regional level. This is accomplished through the rigorous combination of regional or local network solutions utilizing the Solution Independent Exchange Format (SINEX) and a process defined in the densification section.

In the future, the IGS infrastructure could become a valuable asset for support of new ground-based applications and could also contribute to space-based missions in which highly accurate flight and ground differential techniques are required.

ACRONYMS

AE	-	Antenna Electronics
A/D	-	Analog to Digital
AFB	-	Air Force Base
AFI	-	Automatic Fault Indication
AFS	-	Air Force Station
AHRS	-	Attitude and Heading Reference System
AIMS	-	Airspace Traffic Control Radar Beacon System M
A/J	-	Anti-Jamming
AOC	-	Auxiliary Output Chip
A-S	-	Anti-Spoofing
ASIC	-	Application Specific Integrated Circuit
ATE	-	Automatic Test Equipment
BCD	-	Binary Code Decimal
BIH	-	Bureau International de L'Heure
BIPM	-	International Bureau of Weights and Measures
BIT	-	Built-In-Test
BPSK	-	Bi Phase Shift Keying
C/A-code	-	Coarse/Acquisition-Code
CADC	-	Central Air Data Computer
CDMA	-	Code Division Multiplex Access
CDU	-	Control Display Unit
CEP	-	Circular Error Probable
CMOS	-	Complementary Metal Oxide Semiconductor
C/No	-	Carrier to Noise Ratio
CRPA	-	Controlled Radiation Pattern Antenna
CSOC	-	Consolidated Space Operations Center
CW	-	Continuous Wave
DAC	-	Digital to Analog Converter
dB	-	Decibel ($X = 10 \text{ Log}_{10} X$ dB)
DGPS	-	Differential GPS
D-Level	-	Depot Level
DLM	-	Data Loader Module
DLR	-	Data Loader Receptable

DLS - Data Loader System
DMA - Defense Mapping Agency
DOD - Department of Defense
DOP - Dilution of Precision
dRMS - Distance Root Mean Square
DRS - Dead Reckoning System
DT&E - Development Test and Evaluation
ECEF - Earth-Centered-Earth-Fixed
ECP - Engineering Change Proposal
EDM - Electronic Distance Measurement
EFIS - Electronic Flight Instrument System
EM - Electro Magnetic
EMCON - Emission Control
ESGN - Electrically Suspended Gyro Navigator
FAA - Federal Aviation Administration
FMS - Foreign Military Sales
FOM - Figure of Merit
FRPA - Fixed Radiation Pattern Antenna
FRPA-GP - FRPA Ground Plane
GaAs - Gallium Arsenide
GDOP - Geometric Dilution of Precision
GMT - Greenwich Mean Time
GPS - Global Positioning System
HDOP - Horizontal Dilution of Precision
HOW - Hand Over Word
HSI - Horizontal Situation Indicator
HV - Host Vehicle
HQ USAF - Headquarters US Air Force
ICD - Interface Control Document
ICS - Initial Control System
IF - Intermediate Frequency
IFF - Identification Friend or Foe
I-Level - Intermediate Level
ILS - Instrument Landing System
INS - Inertial Navigation System
ION - Institute of Navigation
IOT&E - Initial Operational Test and Evaluation
IP - Instrumentation Port
ITS - Intermediate Level Test Set
JPO - Joint Program Office
J/S - Jamming to Signal Ration
JTIDS - Joint Tactical Information Distribution System

L_1 - GPS primary frequency, 1575.42 MHz
L_2 - GPS secondary frequency, 1227.6 MHz
LEP - Linear Error Probable
LRIP - Low Rate Initial Production
LRU - Line Replaceable Unit
LO - Local Oscillator
mB - Millibar
MCS - Master Control Station
MCT - Mean Corrective Maintenance Time
MHz - Megahertz (10^6 Hz)
MLV - Medium Launch Vehicle
MmaxCT - Maximum Corrective Maintenance Time
MOU - Memorandum of Understanding
M/S - Meters per Second
MSL - Mean Sea Level
MTBF - Mean Time between Failures
MTBM - Mean Time between Maintenance
N/A - Not Applicable
NAV-msg - Navigation Message
NOSC - Naval Ocean Systems Center
NRL - Naval Research Laboratory
NS - Nanosecond (10^{-9} second)
NSA - National Security Agency
NTDS - Navy Tactical Data System
NTS - Navigation Technology Satellite
OBS - Omni Bearing Select
OCS - Operational Control System
O-Level - Organization Level
OTHT - Over the Horizon Targeting
PC - Personal Computer
P-Code - Precise Code
PDOP - Position Dilution of Precision
PLSS - Precision Location Strike System
P^3I - Pre Planned Product Improvement
PPM - Parts Per Million (10^{-6})
PPS - Precise Positioning Service
PPS-SM - PPS Security Module
PRN - Pseudo Random Noise
PTTI - Precise Time and Time Interval
PVT - Position Velocity and Time
RAM - Reliability and Maintainability
RCVR - Receiver

RF - Radio Frequency
RMS - Root Mean Square
RSS - Root Sum Square
RT - Remote Terminal
RTCA - Radio Technical Commission for Aeronautic
RTCM - Ratio Technical Commission for Maritime Services
S/A - Selective Availability
SAMSO - Space and Missile Systems Organization
SBB - Smart Buffer Box
SC - Special Committee
SEP - Spherical Error Probable
SI - International System of Units
SIL - System Integration Laboratory
SINS - Ship borne INS
SPS - Standard Positioning Service
SRU - Shop Replacable Unit
STDCDU - Standard CDU
TACAN - Tactical Air Navigation
TAI - International Atomic Time
TBD - To Be Determined
TDOP - Time Dilution of Precision
TFOM - Time Figure of Merit
TTFF - Time to First Fix
UE - User Equipment
UERE - User Equivalent Range Error
UHF - Ultra High Frequency
USA - United States of America
USNO - US Naval Observatory
UT - Universal Time
UTC - Universal Time Coordinated
VDOP - Vertical Dilution of Precision
VHSIC - Very High Speed Integrated Circuit
VLSIC - Very Large Scale Integrated Circuit
VOR - Very High Frequency (VHF) Omni directional Range
WGS-84 - World Geodetic System - 1984
YPG - Yuma Proving Ground
1 PPM - 1 Pulse per Minute
1 PPS - 1 Pulse per Second

GLOSSARY

Accuracy
A measure of how close an estimate of a GPS position is to the true location.

Acquisition Time
The time it takes a GPS receiver to acquire satellite signals and determine the initial position.

Active Antenna
An antenna that amplifies the GPS signal before sending it to the receiver.

Active Leg
The segment of a route currently being traveled. A "segment" is that portion of a route between any two waypoints in the route.

Almanac Data
Information transmitted by each satellite on the orbits and state (health) of every satellite in the GPS constellation. Almanac data allows the GPS receiver to rapidly acquire satellites shortly after it is turned on.

Altimeter
An instrument for determining elevation, especially an aneroid barometer used in aircraft that senses pressure changes accompanying changes in altitude.

Ambiguity
The unknown integer number of cycles of the reconstructed carrier phase contained in an unbroken set of measurements from a single satellite passes at a single receiver.

Analog Signal

The principal feature of analog signals is that they are continuous. In contrast, digital signals consist of values measured at discrete intervals.

Anti-Spoofing

Encryption of the P-code to protect the P-signals from being "spoofed" through the transmission of false GPS signals by an adversary.

Atmospheric Propagation Delay

Time delay affecting satellite signals due to troposphere layers of the Earth's atmosphere.

Atomic Clock

A very precise clock that operates using the elements cesium or rubidium. A cesium clock has an error of one second per million years. GPS satellites contain multiple cesium and rubidium clocks.

Attribute

A characteristic which describes a Feature. Attributes can be thought of as questions which are asked about the Feature.

Azimuth

The horizontal direction from one point on the Earth to another, measured clockwise in degrees (0-360) from a north or south reference line. An azimuth is also called a bearing.

Bandwidth

A measure of the width of the spectrum of a signal (frequency domain representation of a signal) expressed in Hertz.

Baseline

The length of the three-dimensional vector between a pair of stations for which simultaneous GPS data has been collected and processed with differential techniques.

Beacon

Stationary transmitter that emits signals in all directions (also called a non-directional beacon). In DGPS, the beacon transmitter also broadcasts pseudo range correction data to nearby GPS receivers for greater accuracy.

Bearing

The compass direction from a position to a destination, measured to the nearest degree (also call an azimuth). In a GPS receiver, bearing usually refers to the direction to a waypoint.

Beat Frequency

Either of the two additional frequencies obtained when signals of two frequencies are mixed. The beat frequencies are equal to the sum or difference of the original frequencies, respectively.

Binary Biphase Modulation

Phase changes of either 0° or 180° (to represent binary 0 or 1, respectively) on a constant frequency carrier. These can be modeled by

$$y = A \cos (wt + p),$$

Where the amplitude function A is a sequence of +1 and -1 values (to represent 0° and 180° phase changes respectively). GPS signals are biphaseModulated.

C/A Code

The Coarse/Acquisition GPS code modulated on the GPS L1 signal. This code is a sequence of 1023 pseudorandom binary biphase modulations on the GPS carrier at a chipping rate of 1.023 MHz, thus having a code repetition period of one millisecond.

Cartesian Coordinates

The coordinates of a point in space given in three mutually perpendicular dimensions (x, y, z) from the origin.

Carrier

A radio wave having at least one characteristic (e.g., frequency, amplitude, phase) which may be varied from a known reference value by modulation.

Carrier Beat Phase

The phase of the signal which remains when the incoming Doppler-shifted satellite carrier signal is beat (the difference frequency signal is generated) with the nominally constant reference frequency generated in the receiver.

Carrier Frequency

The frequency of the unmodulated fundamental output of a radio transmitter. The GPS L1 carrier frequency is 1575.42 MHz, the GPS L2 carrier frequency is 1227.60 MHz. The time interval of either a zero or a one in a binary pulse code.

Carrier-aided Tracking

A signal processing technique that uses the GPS carrier signal to achieve an exact lock on the pseudo random code generated by the GPS satellite. Carrier-aided tracking is more accurate than standard C/A Code tracking.

Cartography

The art or technique of making maps or charts. Many GPS receivers have detailed mapping—or cartography—capabilities.

Channel

A channel of a GPS receiver consists of the circuitry necessary to track the signal from a single GPS satellite.

Chip Rate

Number of chips per second (e.g., C/A code: 1.023*106 cps)

Code

A system used for communication in which arbitrarily chosen strings of zeros and ones are assigned definite meanings.

Compacted Data

Raw data compacted over a specified time interval (compaction time) into one single observable (measurement) for recording.

Conformal Projection

A map projection that preserves angles on the ellipsoid after they have been mapped onto the plane.

Control Segment

Ground-based GPS System equipment operated by the U.S. Government that tracks the satellite signals, determines the orbits of the satellites, and transmits orbit definitions to the memories of the Satellites.

Code Division Multiple Access (CDMA)

A method whereby many radios use the same frequency, but each one has a unique code. GPS uses CDMA techniques with codes for their unique cross-correlation properties.

Clock Bias

The difference between the indicated clock time in the GPS receiver and true universal time (or GPS satellite time).

Clock offset

A constant difference in the time reading between two clocks normally used to indicate a difference between two time zones.

Coarse/Acquisition Code (C/A Code)

The standard positioning signal the GPS satellite transmits to the civilian user. It contains the information the GPS receiver uses to fix its position and time, and is accurate to 100 meters or better.

Cold Start

The power-on sequence where the GPS receiver downloads almanac data before establishing a position fix.

Coordinates

A set of numbers that describes your location on or above the Earth. Coordinates are typically based on latitude/ longitude lines of reference or a global/regional grid projection (e.g., UTM, MGRS, and Maidenhead).

Coordinated Universal Time (UTC)

Replaced Greenwich Mean Time (GMT) as the world standard for time in 1986. UTC uses atomic clock measurements to add or omit leap seconds each year to compensate for changes in the rotation of the Earth.

Course

The direction from the beginning landmark of a course to its destination (measured in degrees, radians, or mils), or the direction from a route waypoint to the next waypoint in the route segment.

Course Deviation Indicator (CDI)

A technique for displaying the amount and direction of cross track error (XTE).

Course Made Good (CMG)

The bearing from the 'active from' position (your starting point) to your present position.

Course over Ground (COG)

Your direction of movement relative to a ground position.

Course to steer

The heading you need to maintain in order to reach a destination.

Course up Orientation

Fixes the GPS receiver's map display so the direction of navigation is always "up."

Cross track Error (XTE/XTK)

The distance you are off the desired course in either direction.

Cutoff Angle

The minimum elevation angle below which no more GPS satellites are tracked by the sensor.

Cycle Slip

A loss of continuity in the measured carrier beat phase which results from a temporary loss of lock on a GPS satellite.

Data Message

A message included in the GPS signal that reports the satellite's location, clock corrections, and health. Included is rough information on the status of other Satellites in the constellation.

Datum

A math model which depicts a part of the surface of the Earth. Latitude and longitude lines on a paper map are referenced to a specific map datum. The map datum selected on a GPS receiver needs to match the datum listed on the corresponding paper map in order for position readings to match.

Desired Track (DTK)

The compass course between the "from" and "to" waypoints.

Differential GPS

The term commonly used for a GPS system that utilizes differential code corrections to achieve an enhanced positioning accuracy of around 0.5 - 5m.

Deflection of the Vertical

The angle between the normal to the ellipsoid and the vertical (true plumb line). It is usually resolved into a component in the meridian and a component perpendicular to the meridian.

Delay Lock

The technique whereby the received code (generated by the satellite clock) is compared with the internal code (generated by the receiver clock) and the latter shifted in time until the two codes match.

Differential Correction

The technique of comparing GPS data collected in the field to GPS data collected at a known point. By collecting GPS data at a known point, a correction factor can be determined and applied to the field GPS data.

Differential GPS (DGPS)

An extension of the GPS system that uses land-based radio beacons to transmit position corrections to GPS receivers. DGPS reduces the effect of selective availability, propagation delay, etc. and can improve position accuracy to better than 10 meter.

Differenced Measurements

GPS measurements can be differenced across receivers, across satellites and across time. Although many combinations are possible, the present convention for GPS phase measurement differencing is to perform the differences in the above order: first across receivers, second across satellites and third across time.

A single difference measurement (across receivers) is the instantaneous difference in phase of a received signal, measured by two receivers simultaneously observing one satellite

A double difference measurement (across receivers and satellites) is obtained by differencing the single difference for one satellite with respect to the corresponding single difference for a chosen reference satellite.

A triple difference measurement (across receivers, satellites and time) is the difference between a double difference at one epoch of time and the same double difference at another epoch of time.

Differential Positioning

Determination of relative coordinates between two or more receivers which are simultaneously tracking the same GPS signals.

Dilution of Precision (DOP)

A description of the purely geometrical contribution to the uncertainty in a position fix. The DOP factor indicates the geometrical strength of the satellite constellation at the time of measurement. Standard terms in the case of GPS are

GDOP three positions coordinates plus clock offset

PDOP three coordinates

HDOP two horizontal coordinates

VDOP height only

TDOP clock offset only

HTDOP horizontal position and time

Dilution of Precision (DOP)

A measure of the GPS receiver/satellite geometry. A low DOP value indicates better relative geometry and higher corresponding accuracy. The DOP indicators are GDOP (geometric DOP), PDOP (position DOP), HDOP (horizontal DOP), VDOP (vertical DOP), and TDOP (time clock offset).

Distance

The length (in feet, meters, miles, etc.) between two waypoints or from your current position to a destination waypoint. This length can be measured in straight-line (rhumb line) or great-circle (over the Earth) terms. GPS normally uses great circle calculations for distance and desired track.

DOD

The United States Department of Defense. The DOD manages and controls the Global Positioning System.

Doppler shift

The apparent change in frequency of a received signal due to the rate of change of the range between the transmitter and receiver.

Downlink

A transmission path for the communication of signals and data from a communications satellite or other space vehicle to the Earth.

Dynamic Positioning

The process of collecting GPS data while the GPS antenna is in motion. Often associated with Line or Area Features.

Eccentricity

The ratio of the distance from the centre of an ellipse to its focus to the semi major axis.

$$e = (1 - b2/a2)1$$

Where a and b are the semi major and semi minor axis of the ellipse, respectively.

Electromagnetic Radiation

Electromagnetic radiation is simply energy which travels through space at about 300,000 kilometres per second – the speed of light. We can imagine radiation moving like a wave. The distance between two adjacent wave crests is called a wavelength. The shorter the wavelength, the more energetic the radiation is said to be. Also, the shorter the wavelength, the greater the frequency of the radiation. The highest frequencies in the spectrum of electromagnetic radiation are gamma-rays; the lowest frequencies are radio waves.

Elevation

The distance above or below mean sea level.

Ellipsoid

A geometric surface, all of whose plane sections are either ellipses or circles.

Ellipsoid Height

The vertical distance of a point above the ellipsoid.

Epoch

A particular fixed instant of time used as a reference point on a time scale.

Ephemeris

The predicted changes in the orbit of a satellite that are transmitted to the GPS receiver from the individual satellites.

Ephemeris Data

Current satellite position and timing information transmitted as part of the satellite data message. A set of ephemeris data is valid for several hours.

Ephemeris Errors

Errors which originate in the ephemeris data transmitted by a GPS satellite. Ephemeris errors are removed by differential correction.

Equipotential Surface

A mathematically defined surface where the gravitational potential is the same at any point on that surface. An example of such a surface is the geoid.

Estimated Position Error (EPE)

A measurement of horizontal position error in feet or meters based upon a variety of factors including DOP and satellite signal quality.

Estimated Time Enroute (ETE)

The time it will take to reach your destination (in hours/ minutes or minutes/seconds) based upon your present position, speed, and course.

Estimated Time of Arrival (ETA)

The estimated time you will arrive at a destination.

Flattening

Relating to Ellipsoids.
f = (a-b)/a = 1-(1-e2)1
Where a ... semi major axis
b ... semi minor axis
e ... eccentricity

Frequency

A measure of how frequently a vibration repeats itself (oscillates) or the number of waves passing by in a second. A hertz is the unit of frequency – 1 oscillation per second; a kilohertz (kHz) is 1000 hertz – 1000 oscillations per second.

Fundamental Frequency

The fundamental frequency used in GPS is 10.23 MHz. The carrier frequencies L1 and L2 are integer multiples of the fundamental frequency.
L_1 = 154F = 1575.42 MHz
L_2 = 120F = 1227.60 MHz

Feature

The object which is being mapped for use in a GIS system. Features may be points, lines or areas.

Featuring

The process of collecting GPS and GIS information simultaneously.

Frequency

The number of repetitions per unit time of a complete waveform, as of a radio wave (see L1 and L2 frequencies in this glossary).

Galileo

A satellite radio navigation system initiated by the European Union and developed for non-military applications. The final system will be based on a collection of 30 satellites.

Geocaching

A high-tech version of hide-and-seek. Geocachers seek out hidden treasures utilizing GPS coordinates posted on the Internet by those hiding the cache.

Geocentric

Relating to the centre of the Earth.

Geodesy

The study of the Earth's size and shape

Geodetic Coordinates

Coordinates defining a point with reference to an ellipsoid. Geodetic Coordinates are defined using latitude, longitude and ellipsoidal height or using Cartesian coordinates.

Geodetic Datum

A mathematical model designed to best fit part or all of the geoids. It is defined by an ellipsoid and the relationship between the ellipsoid and a point on the topographic surface established as the origin of datum. This relationship can be defined by six quantities, generally (but not necessarily) the geodetic latitude, longitude, and the height of the origin, the two components of the deflection of the vertical at the origin, and the geodetic azimuth of a line from the origin to some other point.

Geoids

The particular equipotent surface which coincides with mean sea level, and which may be imagined to extend through the continents. This surface is everywhere perpendicular to the direction of the force of gravity.

Geoids Separation

The distance from the surface of the reference ellipsoid to the geoids measured outward along the normal to the ellipsoid.

Geodetic Datum

A math model representing the size and shape of the Earth (or a portion of it).

Geographic Information System (GIS)

A computer system or software capable of assembling, storing, manipulating, and displaying geographically referenced information (i.e., data identified according to their location). In practical use, GIS often refers to the computer system, software, and the data collection equipment, personnel, and actual data.

Geosynchronous Orbit

A specific orbit around where a satellite rotates around the Earth at the same rotational speed as the Earth. A satellite rotating in geosynchronous orbit appears to remain stationary when viewed from a point on or near the equator. It is also referred to as a geostationary orbit.

Global Positioning System (GPS)

A global navigation system based on 24 or more satellites orbiting the Earth at an altitude of 12,000 statue miles and providing very precise, worldwide positioning and navigation information 24 hours a day, in any weather. Also called the NAVSTAR system.

GPS Time

A continuous time system based on the Coordinated Universal Time (UTC) from 6th January 1980.

GLONASS

Global Navigation Satellite System, operated by the Russian Federation Ministry of Defense. When completed, it will have a constellation of 24 satellites, and is intended to service maritime and aviation users throughout the world.

Greenwich Mean Time (GMT)

The mean solar time for Greenwich, England, which is located on the Prime Meridian (zero longitude). Based on the rotation of the Earth, GMT is used as the basis for calculating standard time throughout most of the world.

Grid

A pattern of regularly spaced horizontal and vertical lines forming square zones on a map used as a reference for establishing points. Grid examples are UTM, MGRS, and Maidenhead.

Great Circle Course

Term used in navigation. Shortest connection between two points.

Graticule

A plane grid representing the lines of Latitude and Longitude of an ellipsoid.

Gravitational Constant

The proportionality constant in Newton's Law of gravitation. G = 6.672 * 10-11 m3s-2kg-.

Heading

The direction in which a vehicle is moving. For air and sea operations, this may differ from actual Course over Ground (COG) due to winds, currents, etc.

Healthy

A term used when an orbiting GPS satellite is suitable for use. "State" is also used to refer to satellite health.

Inclination

The angle between the orbital plane of an object and some reference plane (e.g., equatorial plane).

Input/Output (I/O)

The two-way transfer of GPS information with another device, such as a nav plotter, autopilot, or another GPS unit.

Initialization

The first time a GPS receiver orients itself to its current location and collects almanac data. After initialization has occurred, the receiver remembers its location and acquires a position more quickly because it knows which satellites to look for.

Ionosphere

A region of the Earth's atmosphere where ionization caused by incoming solar radiation affects the transmission of GPS radio waves. It extends from a height of 50 kilometers (30 miles) to 400 kilometers (250 miles) above the surface.

Ionospheric Delay

A wave propagating through the ionosphere (which is a non-homogeneous and dispersive medium) experiences delay. Phase delay depends on electron content and affects carrier signals. Group delay depends on dispersion in the ionosphere as well, and affects.

Signal modulation (codes). The phase and group delay are of the same magnitude but opposite sign.

Invert Route

To display and navigate a route from end to beginning for purposes of returning to the route's starting point.

Kinematic Positioning

Determination of a time series of sets of coordinates for a moving receiver, each set of coordinates being determined from a single data sample, and usually computed in real time.

Keplerian Orbital Elements

Allow description of any astronomical orbit:
a: semi major axis
e: eccentricity
W: argument of perigee
W: right ascension of ascending node
I: inclination
N: true anomaly

Lambert Projection

A conformal conic map projection that projects an ellipsoid onto a plane surface by placing a cone over the sphere.

Latitude

The angle between the ellipsoidal normal and the equatorial plane. Latitude is zero on the equator and 90° at the poles.

L1 Frequency

One of the two radio frequencies transmitted by the GPS satellites. This frequency carries the Coarse Acquisition Code (C/A code), P-Code, and the nav message, and is transmitted on a frequency of 1575.42 MHz.

L2 Frequency

One of the two radio frequencies transmitted by the GPS satellites. This frequency carries only the P-Code, and is transmitted on a frequency of 1227.6 MHz.

L Band

The radio frequencies that extend from 390 MHz to 1550 MHz. The GPS carrier frequencies are in the L band (1227.6 MHz and 1575.42 MHz).

Least Squares Estimation

The process of estimating unknown parameters by minimizing the sum of the squares of measurement residuals.

Local Ellipsoid

An Ellipsoid that has been defined for and fits a specific portion of the Earth. Local ellipsoids usually fit single or groups of countries.

Local Time

Local time equals to GMT time + time zone.

Longitude

Longitude is the angle between the meridian ellipse which passes through Greenwich and the meridian ellipse containing the point in question. Thus, Latitude is 0° at Greenwich and then measured either eastward through 360° or eastward 180° and westward 180°.

Latitude

A position's distance north or south of the equator, measured by degrees from zero to 90. One minute of latitude equals one nautical mile.

Leg (Route)

A portion of a route consisting of a starting (from) waypoint and a destination (to) waypoint. A route that is comprised of waypoints A, B, C, and D would contain three legs. The route legs would be from A to B, from B to C, and from C to D.

Lithium Battery

A soft, silvery, highly reactive metallic element that is used in batteries where weight and cold weather conditions are concerns.

Line Of Sight (LOS) Propagation

Of an electromagnetic wave, propagation in which the direct transmission path from the transmitter to the receiver is unobstructed. The need for LOS propagation is most critical at GPS frequencies.

Liquid Crystal Display (LCD)

A display circuit characterized by a liquid crystal element sandwiched between two glass panels. Characters are produced by applying an electric field to liquid crystal molecules and arranging them to act as light filters.

Local Area Augmentation System (LAAS)

The implementation of ground-based DGPS to support aircraft landings in a local area (20-mile range).

Longitude

The distance east or west of the prime meridian (measured in degrees). The prime meridian runs from the north to South Pole through Greenwich, England.

LORAN

Loran, which stands for LOng RAnge Navigation, is a grid of radio waves in many areas of the globe that allows accurate position plotting. Loran transmitting stations around the globe continually transmit 100 kHz radio signals. Special shipboard Loran receivers interpret these signals and provide readings that correspond to a grid overprinted on nautical charts. By comparing signals from two different stations, the mariner uses the grid to determine the position of the vessel.

Magnetic North

Represents the direction of the north magnetic pole from the observer's position. The direction a compass points.

Magnetic Variation

In navigation, at a given place and time, the horizontal angle (or difference) between true north and magnetic north. Magnetic variation is measured east or west of true north.

Map Display

A graphic representation of a geographic area and its features.

Mean Sea Level

The average level of the ocean's surface, as measured by the level halfway between mean high and low tide. Used as a standard in determining land elevation or sea depths.

Meridian

An imaginary line joining north to South Pole and passing through the equator at

Multipath

The interference to a signal that has reached the receiver antenna by multiple paths; usually caused by the signal being bounced or reflected. Signals from satellites low on the horizon will have high multipath error. Receivers that can be configured to "mask out" signals from such satellites can help minimize multi-path.

Multipath Error

An error caused when a satellite signal reaches the GPS receiver antenna by more than one path. Usually caused by one or more paths being bounced or reflected. The TV equivalent of multipath is "ghosting."

Multiplexing Receiver

A GPS receiver that switches at a very rapid rate between satellites being tracked. Typically, multiplexing receivers require more time for satellite acquisition and are not as accurate as parallel channel receivers. Multiplexing receivers are also more prone to lose a satellite fix in dense woods than parallel channel GPS receivers.

Nautical Mile

A unit of length used in sea and air navigation, based on the length of one minute of arc of a great circle, especially an international and U.S. unit equal to 1,852 meters (about 6,076 feet).

Navigation

The act of determining the course or heading of movement. This movement could be for a plane, ship, automobile, person on foot, or any other similar means.

Navigation Message

The message transmitted by each GPS satellite containing system time, clock correction parameters, ionospheric delay model parameters, and the satellite's ephemeris data and health. The information is used to process GPS signals to give the user time, position, and velocity. Also known as the data message.

NAVSTAR

The official U.S. Government name given to the GPS satellite system. NAVSTAR is an acronym for NAVigation Satellite Timing and Ranging.

NMEA (National Marine Electronics Association)

A U.S. standards committee that defines data message structure, contents, and protocols to allow the GPS receiver to communicate with other pieces of electronic equipment aboard ships.

NMEA Standard

A NMEA standard defines an electrical interface and data protocol for communications between marine instrumentation.

North Up Orientation

Fixes the GPS receiver's map display so north is always fixed at the top of the screen.

Observing Session

A period of time over which GPS data is collected simultaneously by two or more receivers.

Orthometric Height

The distance of a point above the geoids measured along the plumb line through the point (height above mean sea level).

Parallel Channel Receiver

A continuous tracking receiver using multiple receiver circuits to track more than one satellite simultaneously.

P-Code

The precise code of the GPS signal typically used only by the U.S. military. It is encrypted and reset every seven days to prevent use from unauthorized persons.

Pixel

A single display element on an LCD screens. The more pixels, the higher the resolution and definition.

Point Positioning

The independent reduction of observations made by a particular receiver using the pseudo range information broadcast from the satellites.

Post Processing

The process of computing positions in non-real-time, using data previously collected by GPS receivers.

Precise Positioning Service (PPS)

The highest level of point positioning accuracy provided by GPS. It is based on the dual-frequency P - code.

Position

An exact, unique location based on a geographic coordinate system.

Position Fix

The GPS receiver's computed position coordinates.

Position Format

The way in which the GPS receiver's position will be displayed on the screen. Commonly displayed as latitude/longitude in degrees and minutes, with options for degrees, minutes and seconds, degrees only, or one of several grid formats.

Prime Meridian

The zero meridian, used as a reference line from which longitude east and west is measured. It passes through Greenwich, England.

Pseudolite

The ground-based differential GPS station which transmits a signal with a structure similar to that of an actual GPS satellite.

Pseudorandom Noise (PRN) code

Any group of binary sequences that appear to be randomly distributed like noise, but which can be exactly distributed. The most important property of PRN codes is that the sequence has a minimum autocorrelation value, except at zero lag.

Pseudo range

A measure of the apparent signal propagation time from the satellite to the receiver antenna, scaled into distance by the speed of light. The apparent propagation time is the difference between the time of signal reception (measured in the receiver time frame) and the time of emission (measured in the satellite time frame). Pseudorange differs from the actual range by the influence of satellite and user clock.

Pseudo-Random Code

The identifying signature signal transmitted by each GPS satellite and mirrored by the GPS receiver in order to separate and retrieve the signal from background noise.

Quadrifilar Helix Antenna

A type of GPS antenna in which four spiraling elements form the receiving surface of the antenna. For GPS use, quadrifilar antennas are typically half-wavelength or quarter-wavelength size and encased in a plastic cylinder for durability.

Radio Technical Commission for Maritime Services (RTCM) Special Committee 104

A committee created for the purposes of establishing standards and guidance for interfacing between radio beacon-based data links and GPS receivers, and to provide standards for ground-based differential GPS stations.

RAIM

Receiver Autonomous Integrity Monitoring; A GPS receiver system that would allow the receiver to detect incorrect signals being transmitted by the satellites by comparing solutions with different sets of satellites.

Range

Term used in Navigation for the length of the trajectory between two points. The trajectory is normally the great circle or the rhumb line.

Rapid Static Survey

Term used in connection with the GPS System for static survey with short observation times. This type of survey is made possible by the fast ambiguity approach that is resident in the SKI software.

Raw Data

Original GPS data taken and recorded by a receiver.

Real Time Kinematic

A term used to describe the procedure of resolving the phase ambiguity at the GPS receiver so that the need for post-processing is removed.

Receiver Channel

The radio frequency and digital hard- ware and the software in a GPS receiver, required tracking the signal from one GPS satellite at one of the two GPS carrier frequencies.

Reconstructed Carrier Phase

The difference between the phase of the incoming Doppler-shifted GPS carrier and the phase of a nominally-constant reference frequency generated in the receiver.

Residual

A quality indicator for a GPS position that is determined during the differential correction process. Indicates uncorrectable error. High residuals are not desirable.

Rhumb Line

Term used in navigation. Trajectory between two points with constant bearing.

RINEX

Receiver INdependent EXchange format. A set of standard definitions and formats to promote the free exchange of GPS data.

Route

A group of waypoints entered into the GPS receiver in the sequence you desire to navigate them.

Satellite Constellation

The group of GPS satellites from which data is used to determine a position. Satellite Configuration.

The state of the satellite constellation at a specific time, relative to a specific user or set of users.

Search the Sky

A message shown when a GPS receiver is gathering satellite almanac data. This data tells the GPS receiver where to look for each GPS satellite.

Serial Communication

The sequential transmission of the signal elements of a group representing a character or other entity of data. The characters are transmitted in a sequence over a single line, rather than simultaneously over two or more lines, as in parallel transmission. The sequential elements may be transmitted with or without interruption.

Selective Availability (SA)

The random error, which the government can intentionally add to GPS signals, so that their accuracy for civilian use is degraded. SA is not currently in use.

Sidereal Day

Time interval between two successive upper transits of the vernal equinox. A location where a receiver has been setup to determine coordinates.

Solar Day

Time interval between two successive upper transits of the Sun.

SONAR

A system using transmitted and reflected underwater sound waves to detect and locate submerged objects or measure the distance to the floor of a body of water.

Space Segment

The satellite portion of the complete GPS system.

Speed Over Ground (SOG)

The actual speed the GPS unit is moving over the ground. This may differ from airspeed or nautical speed due to such things as head winds or sea conditions. For example, a plane that is going 120 knots into a 10-knot head wind will have a SOG of 110 knots.

Spread Spectrum

The received GPS signal is wide bandwidth and low power. The L-band signal is modulated with a pseudo-random noise code to spread the signal energy over a much wider bandwidth than the signal information bandwidth. This provides the ability to receive all satellites unambiguously and to give some resistance to noise and multipath.

Squared Reception Mode

A method used for tracking GPS L2 signals which doubles the carrier frequency and does not use the P-code.

Squaring-type Channel

A GPS receiver channel which multiplies the received signal by itself to obtain a second harmonic of the carrier which does not contain the code modulation.

Standard Positioning Service (SPS)

Level of point positioning accuracy provided by GPS based on the single frequency C/A - code.

Statute Mile

A unit of length equal to 5,280 feet or 1,760 yards (1,609 meters) used in the U.S. and some other English-speaking countries.

Static Positioning

The process of averaging GPS positions taken successively over a period of time with a stationary antenna to increase accuracy.

Static Survey

The expression static survey is used in connection with GPS for all non-kinematic survey applications. This includes the following operation modes:
. Static survey
. Rapid static survey

Stop & Go Survey

The term Stop & Go survey is used in connection with GPS for a special kind of kinematic survey. After initialization (determination of ambiguities) on the first site, the roving receiver has to be moved between the other sites without loosing lock to the satellite signal. Only a few epochs are then necessary on these sites to get a solution with survey accuracy. Once loss of lock occurred, a new initialization has to be done.

Straight-Line Navigation

The act of going from one waypoint to another in the most direct line and with no turns.

Time To First Fix (TTFF)

If you have not used your GPS unit for several months, the almanac data for the satellites may be out of date. The unit is capable of recollecting this information on its own, but the process can take several minutes. Time to First Fix (TTFF) is the time it takes a GPS receiver to find satellites after the user first turns it on (when the GPS receiver has lost memory or has been moved over 300 miles from its last location).

Time Zone

Time zone = Local Time - Greenwich Mean Time (GMT). Note that Greenwich Mean Time is approximately equal to GPS time.

Topography

The form of the land of a particular region.

Track Up Orientation

Fixes the GPS receiver's map display so the current track heading is at the top of the screen.

Track (TRK)

Your current direction of travel relative to a ground position (same as Course Over Ground).

Transducer

A device, much like a microphone, that converts input energy of one form into output energy of another. Fishfinders separate and enhance the information received from a transducer to show underwater objects.

Transformation

The process of transforming coordinates from one system to another.

Transit

The predecessor to GPS. A satellite navigation system that was in service from 1967 to 1996.

Transverse Mercator projection

A conformal cylindrical map projection which may be visualized as a cylinder wrapped around the Earth.

Translocation

The method of using simultaneous data from separate stations to determine the relative position of one station with respect to another station. See differential positioning.

Triangulation

A method of determining the location of an unknown point, as in GPS navigation, by using the laws of plane trigonometry.

Troposphere

The lowest region of the atmosphere between the surface of the Earth and the tropopause, characterized by decreasing temperature with increasing altitude. GPS signals travel through the troposphere (and other atmospheric layers).

True North

The direction of the North Pole from your current position. Magnetic compasses indicate north differently due to the variation between true north and magnetic north. A GPS receiver can display headings referenced to true north or magnetic north.

Turn (TRN)

The degrees which must be added to or subtracted from the current heading to reach the course to the intended waypoint.

Universal Time

Local solar mean time at Greenwich Meridian

Universal Transverse Mercator (UTM)

A nearly worldwide coordinate projection system using north and east distance measurements from reference point(s). UTM is the primary coordinate system used on U.S. Geological Survey topographic maps.

Uplink

A transmission path by which radio or other signals are sent from the ground to an aircraft or a communications satellite.

User Equivalent Range Error (UERE)

The contribution to the range measurement error from an individual error source, converted into range units, assuming that error source is uncorrelated with all other error sources.

User Interface

The way in which information is exchanged between the GPS receiver and the user. This takes place through the screen display and buttons on the unit.

User Segment

The segment of the complete GPS system that includes the GPS receiver and operator.

Value

Descriptive information about a Feature. Values can be thought of as the answers to the questions posed by Attributes.

Velocity Made Good (VMG)
The rate of closure to a destination based upon your current speed and course.

Waterproof
Most Garmin GPS units are waterproof in accordance with IEC 529 IPX7. IEC 529 is a European system of test specification standards for classifying the degrees of protection provided by the enclosures of electrical equipment. An IPX7 designation means the GPS case can withstand accidental immersion in one meter of water for up to 30 minutes. An IPX8 designation is for continuous underwater use.

Wavelength
The distance between points of corresponding phase of two consecutive cycles of a wave.

Waypoints
Waypoints are locations or landmarks worth recording and storing in your GPS. These are locations you may later want to return to. They may be check points on a route or significant ground features. (e.g., camp, the truck, a fork in a trail, or a favorite fishing spot). Waypoints may be defined and stored in the unit manually by taking coordinates for the waypoint from a map or other reference. This can be done before ever leaving home. Or more usually, waypoints may be entered directly by taking a reading with the unit at the location itself, giving it a name, and then saving the point. Waypoints may also be put into the unit by referencing another waypoint already stored, giving the reference waypoint, and entering the distance and compass bearing to the new waypoint.

Wide Area Augmentation System (WAAS)
A system of satellites and ground stations that provide GPS signal corrections for better position accuracy. A WAAS-capable receiver can give you a position accuracy of better

than three meters, 95 percent of the time. (At this time, the system is still in the development stage and is not fully operational.) WAAS consists of approximately 25 ground reference stations positioned across the United States that monitor GPS satellite data. Two master stations, located on either coast, collect data from the reference stations and create a GPS correction message.

WGS-84

World Geodetic System, 1984. The primary map datum used by GPS. Secondary datums are computed as differences from the WGS 84 standard.

Y-Code

The encrypted P-Code.

Zenith Angle

Vertical angle with 0° on the horizon and 90° directly overhead.

REFERENCES

Aerospace, (1968). *The 1968 Aerospace Year Book*. Washington D.C.: Books, Inc.

Agrawal, N. K., (2004). *Essentials of GPS*. Hydrabad: Spatial Networks Pvt. Ltd.

Buckeye, K.R., (2004). GIS - GPS for Distance - Based User Fee System, *GIS Development*, 8 (11), pp. 34-37.

Burchell, Wade, Lt. Cmdr, (1997). Reserve Helos Get Magic *Lantern Naval Institute Proceedings*, pp. 74-75,

Curtis Peebles, (1987). *Guardians: Strategic Reconnaissance Satellites*, Novato, CA: Presidio.

Galileo (2005). Galileo Home Page at http:\\Galileo-pgm.org

Ganesh, A. and Narayanakumar R., (2003). *Principles of GPS*. Tiruchirappalli: Bharathidasan University.

Gartner, George, V. Radoczky and G. Retscher, (2005). Location Technologies for Pedestrian Navigation, *GIS Development*, 9 (4), pp. 22-31.

GLONASS, (2005) GLONASS Home Page at http:\\rssi.ru/SFCSIC/SFCSIC_main.html

Gouzhva, Yuri G. and Arvid G. Guevorkyan, Arkady B. Bassevich, Pyotr P. Bogdanov, (1992). High-Precision Time and Frequency Dissemination With Glonass, *GPS World*, 3, (7), pp. 40-49.

GPS, (2005). GPS Home Page at US Coast Guard www.navcen.useg.gov./default.htm

GPS Word, (1992). Protesters invade Rockwell Plant, Damage GPS Satellite, *GPS World*, 3 (6), p. 12.

GPS Word, (1992). Satellite Set Healthy, Launch Planned for Mid-June, *GPS World,* 3 (6), p.12.

GPS Word, (1993). FAA Approves GPS for IFR Operations, *GPS World,* 4, (7), p.18

He, Xiufeng, W. Sang, Y. Chen and X. Ding (2005) *Steep* – Slope Monitoring. GPS Multiple Antenna System at Xiaowan Dam, *GPS World,* 16 (11), pp. 20-25.

Higbie, Paul R. and Norman K. Blocker. (1994). Detecting Nuclear Detonations with GPS, *GPS World,* 5 (2), pp.48-50.

Hofmann-Wellenhof, B. Lichtenegger, H and Collins.J, (2001). *GPS Theory and Practice,* New York: Springer Wien.

Ivan A. Getting. (1993). The Global Positioning System, *IEEE Spectrum,* pp.36-47.

Jethwani, Hiresh, (2004). GPS Navigation in Indian Automobiles, *GIS Development,* 8 (11) p. 44.

Johnson, Nicolas, L., (1994). GLONASS Spacecraft, *GPS World,* 5 (11), pp. 51-58.

Leica (2003). *GPS Basics: Introduction to GPS, Leica Version 1.0.*

Leick Alfred, (2004). *GPS Satellite Surveying, Third Edition,* New Jersey: John Wiley and Sons Inc.

Kaplan E.D. (Ed). (1996). *Undernawy GPS Principles an Application.* Norwood. MA: Artech House.

Kennedy, Michael (2002). *The Global Positioning System and GIS: An Introduction,* London : Taylor and Francis.

Kraak, Menno-Jan and Allan Brown, (2001). *Web Cartography.* London : Taylor and Francis.

Parkinson, Bradford, W., (1994). GPS Eyewitness: The Early Years. *GPS World,* 5 (9), pp. 32-45.

Schiestl, Andrew J., (2004). GPS and Alarka Rail Safety. *GPS World,* 15 (3), pp. 14-19.

Stine, G. Harry (1971). *ICBM.* New York: Orion Books.

Syed, Salmand and E. Cannon (2005). Map-Added GPS Navigation, *GPS World*, 16 (11), pp. 39-44.

Tsui, James Bao-Yen, (2005). *Fundamentals of Global Positioning System Receivers. A Software Approach*, Texas: Wiley.

Whitford, Marty, (2005) Hurricane Hunters : GPS Dropsondes Trace Katrina's Course, *GPS World*, 16 (11), pp. 20-22.

Worthman, Ernest, (1997). *Global Positioning Systems - Technology into the 21st Century, RF Design*, pp. 44-54.

Wysocki, Joseph, Lt. Col., (1991). GPS and Selective Availability - The Military Perspective. *GPS World*, 2, (7), pp. 38-44.

Xie, Yicheen, D. Goby, R. Raymond, R. Pizzi, (2004). Going Beyard Automatic Vehicle Location, *GIS Development*, 8 (11), pp. 30-33.

Web Resources

http://www.aero.org/education/index.html

http://www.buzzle.com/editorials/9-10-2004-115.asp?viewPage=2

http://www.colorado.edu/geography/gcraft/contents.html

http://www.colorado.edu/engineering/ASEN/asen5090/GPS3.sm.jpg

http://www.gisdevelopment.net/gps

http://igscb.jpl.nasa.gov/overview.html

http://www.gmat.unsw.edu.au/snap/gps/gps_survey/chap2/211.htm

http://www.gmat.unsw.edu.au/snap/gps/gps_survey/chap3/311.htm

http://www.gps-practice-and-fun.com/personal-location-tracking.html

http://hyperphysics.phy-astr.gsu.edu/hbase/gpsrec.html

http://www.montana.edu/places/gps/understd.html

http://www.navcen.uscg.mil/loran/loranff.htm

http://www.navcen.uscg.mil/omega/omegaff.htm

http://www.navcen.uscg.mil/gps/gpsff.htm

http://www.palowireless.com

http://science.howstuffworks.com/nuclear2.htm

http://www.trimble.com/gps/how.html

INDEX

D

E

F

G

—•—•—

About the Authors

Dr. A. Ganesh : is a renowned researcher in the field of water resources evaluation and management. His significant contributions to water resources using remote sensing, GIS and GPS can hardly be over emphasized. His findings have helped to map the surplus and deficit areas of water in various river basins of Tamil Nadu, India to analyze the surface and ground water resources and to evolve better water management practices. His academic visits include various universities and research institutions. He has to his credit fifty research papers and three books. His two-decade academic career has earned him many laurels. His reputation rests upon his research programs with various agencies. As a true scholar, he holds teaching and research as the two sides of the same coin. He has been guiding five PhD scholars.

Having got Graduate Degree in Geography from Government Arts College, Coimbatore in the year 1975, he studied further and obtained Post-Graduate Degree from the Presidency College, Chennai in 1977. He earned his doctorate in Geography from the University of Madras, Chennai in 1985 for his work on Geographical Appraisal of Water Resources in Upper Vaigai basin, Tamil Nadu, Southern India under the guidance of Dr Vasantha Vishwanath, an eminent scholar in the field of Regional Climatology. During this period he got training in application of aerial and satellite remote sensing for resources evaluation at Survey of India, Bangalore and Indian Institute of Remote Sensing, Dehra Dun.

After a brief stay as a CSIR Post Doctoral Fellow, he moved over to School of Earth Sciences, Bharathidasan University, and Tiruchirappalli in 1987 as a Lecturer. Besides the concern on water resources, he admixed remote sensing and GIS techniques for a number of environmental problems. During May 1999, he has attended a training programme on Remote Sensing for Watershed Management sponsored by NASDA Japan conducted by the GIS Application Centre, Asian Institute of Technology, Bangkok, Thailand. He is life member of a number of professional societies related to Geography, Remote Sensing and Geomatics and also Chairman of Board of Studies in Geography and Remote Sensing in various universities and colleges.

Dr. A. Ganesh is right now Professor and Head of the Department of Geography, School of Geosciences, Bharathidasan University, Tiruchirappalli, India. He is engaged in conducting two programmes: M Tech in GIS Applications and UGC sponsored Innovative Programme on Post Graduate Diploma in Geomatics.

R. Narayanakumar : is a Guest Faculty in the Department of Geography, School of Geosciences, Bharathidasan University, Tiruchirappalli. He has vast experience both in commercial as well as academic fields of GIS and GPS. He pursued his M.Sc., in Environmental Remote Sensing and Cartography at the Madurai Kamaraj University, Madurai and M.Phil., in Environmental Science and Engineering from Pondicherry Central University, Pondicherry. He has been teaching Principles of GPS and GIS for the students of M.Tech. of GIS Applications and Post Graduate Diploma in Geomatics at the Department of Geography, Bharathidasan University since 2003. He worked as a Spatial Data Executive in DSM Soft Pvt., Ltd, and Tiruchirappalli for about three years. He has published a number of research articles in journals and books.